Mission-Critical Facilities Management

For the Non-Engineer

Eric Woodell, D.Sc.

DEDICATION

To Nadya, Olya and Tasha. They keep me focused on what matters.

CONTENTS

ERIC WOODELL, D.Sc.

ACKNOWLEDGMENTS

Thanks to Rene S. who got me into the data centers. Also Chet Snyder, who gave me some pointers on how to make the book more interesting. I also owe a debt of gratitude to my RMU colleagues, who first put me on the path to understanding the "why" of things instead of merely reacting to the "what."

1 INTRODUCTION

This book is written for the non-engineering manager, supervisor or executive that must manage a mission-critical facility, specifically data centers. But don't worry, the information here is applicable to any facility or building where continuous power, cooling and other services are needed: air traffic control systems, hospitals, FEMA installations and the like. If you have a facility where it is absolutely critical that systems stay functioning, this is the book for you.

If you're in the role of managing such a facility, you'll appreciate the format; it's short, sweet and to the point. You don't have time to read a college text, but instead something that gives you the main points you need to make quick, informed decisions, based on sound data instead of a vendor brochure.

While some engineering theory is necessary to explain the nuts and bolts of a mission-critical facility functions, we avoid getting into deep-dive engineering aspects. In other words, we'll cover basic information on what the different systems need to do, the risks and benefits of how different designs of equipment fulfill those needs and what's best for your budget What do you look for if you decide to build a data center? How do you recruit and manage engineering teams to operate those data centers moving forward?

This book does not get into the Information Technology (IT) aspects; there are plenty of books out there on network architectures, server operations and batch processing; this book is about the building infrastructure needed to support those IT goodies, and provide that support reliably.

At the end of this book are several appendixes; two case studies, recent research on the effectiveness of contemporary IT operating models (and how it effects the reliability of the data center) and a complete training program to deploy in the mission-critical facility. This latter piece is always

1

problematic to develop, and training companies always seem to promise a good product but fail to deliver real results. This gives you everything you need to build a proven, meaningful training program from the ground up- with a little massaging to make it match your facility, of course.

WHY DO DATA CENTERS MATTER?

In the business world, some things are quite easy to understand and quantify in terms of value. The net present value of a production line, given the costs of inputs and the sale price of a product, throughput and other factors, will allow you to easily determine this answer. But what happens, when one of your key business assets is not easily quantified? For example, what is the precise value for the Nike "swoosh," or corporate goodwill? This question becomes far more difficult, and involves a lot of educated guesses. The same applies to data centers, and other critical facilities.

A critical facility, for the purposes of this book, is defined as one that has an inordinate impact on business operations and/or profitability, should key infrastructure systems lose power or other support, such as cooling.

At the level of the common man, a hospital is one such example; you don't want the power to fail while a surgeon is performing open-heart surgery on you. The respirator used to oxygenate your lungs, the external pumps which allow blood to keep flowing through your body, even the lights by which the surgeon and his team need, to complete the operation, all rely upon electricity. So if utility power fails, you'd better have sufficient backup power systems to seamlessly transfer to an alternative power source, typically reserve generators.

If you have a small business, with your office in your home, you use a desktop (or laptop) computer. Should your computer fail, the data loss could erase your company's customer-list, billing records and the like. We would naturally try to provide insurance against such events by periodically backing up our data, so that the losses would be minimized. We could go further, and even buy a small uninterruptible power supply (UPS) from a local electronics store, so that we could perform an orderly shutdown on loss of utility power. All of this is equally true for the large companies, as well. This is where the data center comes into play; it's a building with reserve power and cooling, generators, switchgear and other support systems, to protect the computer systems from failure due to a loss of utility services. The goals of the modern data center are exactly the same as the computer system would be for your home office- but on a massive scale, instead of small, single-user system.

Consider the differences in computing needs, from the home-office, to the corporate realm. The corporation must depend on the data center for their survival. This is because Information Technology (IT) has become so prolific, with everything from shipping and receiving functions, internal email and customer-billing functions, that the IT systems are now the life-blood of the large, modern company. But because the IT systems serve thousands (or millions) of users at approximately the same time, the software is correspondingly more complex.

Think of the software as a house, in that there are the foundations (such as operating systems), followed by the walls (software applications), then the roof (customer interfaces and logins) and the furnishings (real-time data and records, which are accessed by the system users). If a data center were to suffer a complete outage of electrical power, you must start rebuilding the entire software 'house,' layer by layer. Only when that is complete, can you reload your latest data, to 'refurnish' the house, so that it can be used again. And rarely are records perfectly saved until the millisecond before the power outage occurred; there will be errors- both in terms of user data, as well as program flaws that will only surface when you attempt to reload the software, as you rebuild the 'house.'

If you've had your home computer fail, you're probably somewhat familiar with the amount of time involved to not only repair the damage (such as a failed hard-drive on your computer, necessitating physical replacement), you're also going to find out how painful it is, to reload software that you had been using, and then attempting to restore all the critical data that you were using so easily before. And you'll discover errors in the data, missing records, lost photos, and so on... It's a painful process, for a simple home computer.

Now magnify that to hundreds or thousands of servers, thousands of software programs and millions of records, and you'll see that it is even more painful for a corporation, requiring hundreds or thousands of man-hours to restore an IT system that has suffered an unplanned power outage.

Of course, during this time, business operations which rely on IT cannot function. Imagine, for example, if an IT-dependent company like Wal-Mart lost their entire IT functionality. The ramifications would be immense, with no bar-code scanners, no electronic payment (credit or debit cards), little or no pricing on the actual products, no ability to monitor inventory, no ability to track the hours of employees, no on-line shopping, and so on... Such an event would literally bring a company to its knees in a matter of hours or days. Such risks are well know:

In a January, 2007 study conducted by the Economist Intelligence Unit, of the U.S. National Archives and Records Administration, their findings indicated that 25 percent of companies that experienced an IT outage lasting from two to six days went bankrupt immediately. That same report also revealed that 93 percent of companies that lost their data center for 10 or more days filed for bankruptcy within a year (Gold, 2007).

Wal-Mart is a company that has heavily invested in their IT capabilities, and utilized it to build extremely efficient supply systems... but they do provide a physical product, have cargo trucks, physical stores and tellers who can literally take cash for purchases. A loss of IT capabilities would cause significant damage if it were of an extended nature, but short-duration outages would be survivable. If, however, the company relies exclusively on their internet presence- examples include E-Bay, Amazon.com and Google- then the loss of their IT presence represents an enormous threat.

There are two separate approaches that the corporation can take, to minimize- or eliminate- these risks.

The first option is to utilize multiple data centers, with full data "mirroring," so that the data is recorded in multiple locations simultaneously. Should a single data center failure occur, the remaining data centers and their IT gear will automatically assume the load, and there will be no loss of data. This approach allows an uninterrupted IT presence for continued business operations and customer service. Where the business relies heavily on internet usage by customers (E-Bay, for example) this is the most effective way to ensure the customer never has a problem. The downside of this approach is that it is, by far, the most expensive solution available, requiring significant investments in many very expensive buildings, redundant computers, servers, storage systems and other associated systems.

The second approach is to build one- or a few- facilities, where the key support systems have sufficient redundancy inside the building to minimize the possibility for unplanned outages. The advantage of this approach is that it reduces the amount of capital outlays to maintain an IT presence. For example, we might have only one or two data centers instead of 100. The disadvantage is that you'll have all of your eggs a very small number of baskets. While you can conceivably have data mirroring between two data centers so that no records will be lost, the limitations (at this time) for mirroring of data requires the geographic distance to be less than 100km, or ~67 miles (the reasons are beyond the scope of this book). This means that a large regional event such as a hurricane or earthquake could therefore threaten both data centers, putting your company at risk.

The other disadvantage with having one or a few data centers, is that an unplanned power outage at one facility can have an inordinate impact on your business, even with redundancy; records will have to be verified correct, and software will have to be checked to make sure that remaining facility assumed the load, and so on. If you have a single data center, this risk is best minimized by adding as much redundancy as feasible, to minimize the risk of a complete failure of power. This will be discussed in-depth, in this book.

Between the two approaches, most companies take a hybrid approach; multiple data centers, each with a significant amount of redundancy and emergency support systems inside.

The approach that your company takes regarding the protection of your IT assets, should be comprehensively studied, and compared against your business model. By doing this, you will be able to best meet your companies needs for IT reliability, against capital and operational expense budgets.

What Is A Data Center? What IS Availability? What Are Tier and Criticality Levels?

A data center is a centralized repository, either physical or virtual, for the storage, management, and dissemination of data and information organized around a particular body of knowledge or pertaining to a particular business (Godinho, 2000). The data center is designed with sufficient protective systems to compensate for service interruptions, and maintain operational readiness for the Information Technology customers.

The high-operational readiness demands are satisfied by:

a. Having a highly redundant electrical system with multiple utility feeds,

b.   Uninterruptible power supplies for momentary power losses and

c. Internal generators for power losses of an extended length of time.

d.   Featuring specialized cooling systems, to remove heat from the computer systems.

e. The cooling systems will also feature several layers of redundancy,

f. Adaptability to the changing needs of the IT environment over time.

g.   Remote monitoring and control capabilities,

h.   Multiple levels of fire protection.

i. Typical fire regulations specify the requirement only for water sprinklers to be utilized.

ii.   In a data center, however, water should be considered the last line of defense, strictly to save the building.

iii.   The first line of defense should include multiple zones of highly sensitive smoke detectors,

iv.   coupled with a gaseous agent that smothers flames-

1.   typical gases include Halon (which is now no longer produced,

but still utilized in older buildings)
2.   FM-200.
i. Using a variety of approaches to achieve high levels of security.
i. Special architectural features to minimize external threats, as well as
ii.   Control access to specific areas, to minimize internal threats.

For the business which relies heavily on IT, large-scale data centers are a logical choice for protecting their core business.

Why does continuous availability matter in a data center?

A building is constructed layer by layer. You have your concrete foundations, then studs for walls, than rafters, a roof, drywall, plumbing and electrical inside. If there is a storm (or a power outage) your house may suffer some damage; an earthquake may break windows or crack the foundations, but the house may still be useable.

Computer applications for things like logistical operations (as exemplified by Wal-Mart), internet commerce (E-Bay and Amazon) and even social media (Twitter and Facebook) requires hardware and software that is built one layer upon the next, much like that house

In the industry, there are a variety of terms to describe the constant, 24x7x365 presence of a data center. Its most commonly referred to as "uptime" or "availability." What it really means is that power and cooling are always available, and that any IT failure is not due to something as silly as a computer getting too hot, or someone accidentally unplugging your server.

Why is there such a big deal made about availability?

In a data center, if all power is lost to IT equipment (commonly referred to as an "unplanned service outage") than a complete "reboot" of the data center will be required. That involves, manually opening all power feeds to IT equipment, restoring main power to the building, and then powering up the IT equipment in a controlled manner. There will be some issues with this operation, because most data centers have some "legacy" systems which have been running for years; if powered off, a certain percentage of them never restart. This will cause some delays in the organized restart of the data center.

After all the equipment has been restarted, IT engineers and staff will then have to reinstall the software layer-by-layer, much like building another house from the ground up. Again, there will be some errors because no installation program is 100% perfect, and the software may require the use of some legacy equipment that has permanently expired during the outage. With that legacy gear not functioning, corrections will have to be made in the software on-the-fly to account for these unexpected changes.

Finally, after the software layers are rebuilt, than software records and data needed for day-to-day operations will need to be reinstalled, via data storage devices (disk storage arrays, magnetic tapes, etc.). It is likely that some data will be lost or corrupted in the restoration process, and the data will need to be rebuilt or recovered- if possible- through some other means.

Going back to our previous example, we were talking about a how a building is constructed. If a house is damaged in the event of a minor storm or loss of utilities, it may still functional. The damage will be obvious, and easily quantified. In a data center, if there is a partial outage of some computer systems (hardware) due to either loss of electrical power or cooling the software can be become corrupted, giving rise to a plethora of unexpected errors. The damage will not be easily detected, and therefore difficult to quantify and correct. Indeed, the process chasing down and fixing the errors caused by a partial outage of a data center is often longer than if the data center suffered a complete failure.

As you can imagine, mobilizing IT personnel to repair such damage takes a certain amount of time; correcting the damage itself can take anywhere from several hours to several days, depending on how robust the backup software is, how frequently data is saved for emergency recover, the in-place processes and procedures to restore a data center from an outage, etc.

What IS the cost associated with a data center outage? I was tasked with answering this exact question for a particular corporation. Company XXX had gross revenues of $11.25B, resulting in a revenue flow rate of approximately $357/second is the result. Thus, the estimated financial impact of a data center failure is as follows:

Revenue Flow Rate	Four Hours	Twelve Hours	One Day	One Week	One Month
$ 356/sec	$5.1M	$15.4M	$30.8M	$216M	$863M

This assumes a cash-flow equalized over the entire year. Obviously, if an outage occurs off-hours (such as a Saturday night), the costs will be less, and one that occurs on a Monday morning will- by several orders of magnitude- be more costly.

Further investigation into the disaster-recovery tests that were subsequently conducted by this client indicated
A.   The recovery time from a complete data center outage to be at least 36 hours, resulting in an estimated cost of $46 million.
B.   Any outage of approximately 30 seconds or longer would require the 36 hours to recover, regardless of the duration of the outage itself.

If you think that estimated cost is outlandish, consider the results of companies participating in the 2001 Cost of Downtime Survey:
• 46% said each hour of downtime would cost their companies up to $50,000,
• 28% said each hour would cost between $51,000 and $250,000,
• 18% said each hour would cost between $251,000 and $1 million,
• 8% said it would cost their companies more than $1M per hour. (http://www.ontrack.com/library/rdr_2003_whitepaper.pdf.)

Clearly, a data center outage can be expensive to an IT-based company. But what are the overall implications for the industry? Why do such things continue to happen?

It has been my observation, that every company increases their reliance on information technology (IT) equipment with the best of intentions.

Generally, the IT presence begins with simple applications that increase worker productivity by reducing errors and paperwork, and decrease the time expended to perform such things as accounts payable operations, payroll and purchase-order generation. With the natural successes that such simple shifts produce (from human-based to IT-based functions), the expected acceleration towards more IT-based functionality occurs, eventually leading to such customer-based luxuries as on-line bill payment, HR-based functions like recruiting and employee services, accounting applications such as SAP, financial functions like stock offers and trades, etc.

Of course, with the trend towards increased IT-based business options, employees and customers are happier. And with each IT success, the trend grows ever more powerful, to shift more work and functionality to information technology.

The required IT equipment is typically located in a specialized building, called a data center, which provides multiple layers of redundancy, to prevent the failure of the IT systems, if an external event occurs. Data centers are built to provide a specific service life (typically, 20-30 years), power availability and heat-removal capabilities. Often, it is physically hardened, to withstand events such as earthquakes and storms or vandalism, and ride through events like failure of regional services, such as water, gas or electricity. The capacity of the data center is determined before the facility is built, comparing forecasted IT growth against the power availability of local utility providers.

But the trend of companies shifting their core business functions to IT is increasing far beyond anyone's expectations, and the data center is evolving into the locus of business continuity, and correspondingly, the Achilles' heel. Regrettably, IT-reliant companies rarely understand that their very survival will come to depend upon the data center.

Naturally, the accelerating shift to IT applications requires more IT hardware, necessitating more electrical power and cooling. Eventually, the inexorable electrical load creep of the IT equipment exceeds the physical limits of the data center; when this point is reached, data center failure is certain. Simply put, when you pull too much electrical current through copper wires, the resulting failures are rarely small or easy to correct.

Tragically, such situations occur quite often, because the situation develops over a long period of time (years to decades), and the complacency and familiarity by the stakeholders tends to force out any opinions which contradict long-set perceptions of stability within the data center. The lack of data center specialists- who can recognize such trends and risks- allows the problem to metastasize.

Usually, executive managers don't realize how perilous the situation is, until after the primary facility which houses the IT equipment has suffered a catastrophic failure, with a resultant loss of business operations for days, if not weeks. And the long-term results are well-documented. In a January, 2007 study conducted by the Economist Intelligence Unit, of the U.S. National Archives and Records Administration, their findings indicated that "...25 percent of companies that experienced an IT outage lasting from two to six days went bankrupt immediately. That same report also revealed that 93 percent of companies that lost their data center for 10 or more days filed for bankruptcy within a year."

As to the possibilities of a data center outage occurring, according to http://www.datacenterknowledge.com/,

"Over the next five years, power failures and limits on power availability will halt data center operations at more than 90% of all companies. It's a fact of life: Outages happen. If you've dodged the bullet thus far, you're part of a fortunate minority. A recent AFCOM survey found that 81 percent of respondent had experienced a failure in the past five years, and 20 percent had been hit with at least five failures. Of those outages, more than 80 percent were due to either external or internal power failures."

Tiers

The typical industry measurement- developed by the Uptime Institute- are "Tier" levels, 1-4. The Uptime Institute bases their Tier rating on a "performance standard," a high-level set of guidelines rather than strict designs of electrical and mechanical systems, as they recognize that there are various ways to achieve a certain goal. This approach is intended to measure the overall performance of the facility based upon the design of its infrastructure, and then using table-top reviews to determine what would be affected if a particular component fails unintentionally. In this way, while we can never get a true "apples-to-apples" comparison, we can look at the overall performance of the facility, and use that as the basis of comparison. Typical Tier designs will be covered in the Electric Plant chapter.

A Tier-I data center is basically an upgraded office area/IT space; a generator to provide backup power during an extended utility outage, an uninterruptible power supply (UPS to absorb short-term power disturbances, and dedicated cooling to IT areas that stay on constantly. The design- due to its limited capabilities and lack of redundancies- will typically require 2 12-hour shutdowns per year, for maintenance, and it is reasonable to expect 1.2 failures per year, on average, or 28.8 hours.

A Tier-II data center has some redundancies built into key systems (though not all systems), reducing the need for planned maintenance shutdowns to 3 12-hour periods every two years, while increasing the reliability expectations to a .92 failures per year, or 22 hours of unplanned downtime.

A Tier-III facility has sufficient redundancies built in, where concurrent maintenance can be performed. That is, the need for planned maintenance shutdowns has been eliminated, and there is enhanced redundancy to reduce the expectation of unplanned outages to a mere 0.067 average unplanned outages per year. In practical terms, that means you should expect to see 1 4-hour unplanned outage during a 42-month window.

Finally, the Tier-IV facility has not only the ability for concurrent maintenance, but is fault tolerant. This key concept means that the facility can survive a worst-case-scenario failure, without that failure cascading down to affect IT loads. It also means the infrastructure is "self-healing," adapting to changing or multiple failures, to maintain operations even as the conditions change within a plant. In some ways, the self-healing aspect can appear the most difficult factor to design into a system.

Engineering firms use failure-analysis software to estimate the probabilities of certain failures occurring. This approach depends on manufacturer estimates for how long a particular component will last in a system (such as a contact within a 4000-amp breaker), assuming certain usage and environmental conditions. It also assumes a confidence-interval (part of statistical process control) to guess how accurate those estimates are versus reality. As you can see, while this approach appears on the surface to be a "scientific" way to build a high-availability data center, there are a lot of assumptions (guesses), which make this modeling approach little more than a sales-tool for engineers trying to sell you the "perfect" design. Indeed, this modeling has resulted in different design engineers to tell me their approach is a "Tier-3.5." If you, as the customer of a design engineering team, are told that your facility with be Tier X.5 or X.2 (or whatever), my advice is to find another design engineering firm, as there is no ".5" of a Tier. Indeed, the design, like a chain, is limited by the weakest link. If your cooling system is Tier-II and your electric plant is Tier-IV, you have a Tier-II facility, with the cooling system being the limiting factor in your plant.

Here is a summary of Tier ratings and expectations:

Tier	Overall availability	# Planned shutdowns required?	Average # of unplanned outages/year	Hours of unplanned downtime per year?
1	99.67%	2 per year, 12-hour duration	1.2	28.8
2	99.75	3 per 2 years	.92	22
3	99.98	0	.067	1 4-hour event/(2.5 years)
4	99.995	0	.033	2 4-hour event/(5 years)

In practical terms, you should logically expect a Tier-I site to have one unplanned outage per year (if not more), a Tier-II should be slightly better. A Tier-III will have very good performance, and be adequate for all but the most stringent requirements, and a Tier-IV facility should provide the best protection for your data and IT operations.

The Telecommunications Industry Association modified the Uptime Institute Tier rating, and developed the TIA-942. While it is not strictly required that TIA-942 is adhered to, it provides guidelines for electrical and mechanical system design to help data center professionals achieve a desired level of reliability.

Criticality Levels

To give a more concrete example of such measurements, we will also look at the Syska-Hennessy Group-developed concept of Criticality Levels, improving upon the Uptime Institute standard, to have ten distinct levels of criticality. The Criticality Levels approach examines not only the design of the data center support systems, but also maintenance and operational concerns, to flesh-out the overall effectiveness of a data center beyond the components it was built with.

Criticality (Avelar, Victor, 2007)	Business Characteristics	Effect on system design
1 *(Lowest)*	• Typically small businesses • Mostly cash-based • Limited online presence • Low dependence on IT • Perceive downtime as a tolerable inconvenience	• Numerous single points of failure in all aspects of design • No generator if UPS has 8 minutes of backup time • Extremely vulnerable to inclement weather conditions • Generally unable to sustain more than a 10 minute power outage
2	• Some amount of online revenue generation • Multiple servers • Phone system vital to business • Dependent on email • Some tolerance to scheduled downtime	• Some redundancy in power and cooling systems • Generator backup • Able to sustain 24 hour power outage • Minimal thought to site selection • Vapor barrier • Formal data room separate from other areas
3	• World-wide presence • Majority of revenue from online business • VoIP phone system • High dependence on	• Two utility paths (active and passive) • Redundant power and cooling systems • Redundant service providers • Able to sustain 72-hour power

	IT	outage
	• High cost of downtime	• Careful site selection planning
	• Highly recognized brand	• One-hour fire rating
		• Allows for concurrent maintenance
4 *(Highest)*	• Multi-million dollar business	• Two independent utility paths
	• Majority of revenues from electronic transactions	• 2N power and cooling systems
		• Able to sustain 96 hour power outage
	• Business model entirely dependent on IT	• Stringent site selection criteria
		• Minimum two-hour fire rating
	• Extremely high cost of downtime	• High level of physical security
		• 24/7 onsite maintenance staff

Key questions can determine the Tier rating you'll need for your data center:

As I'm sure you noticed in the charts above, the fundamental question to be answered is the cost of downtime. If, as mentioned before, you have a relatively low cost of downtime, then a Tier-I or Tier-II approach is certainly acceptable for your needs. A university research data center, for example, probably won't require a 24x7xforever approach.

Can the failure of your computer system endanger the lives of people (such as hospitals or air-traffic control systems at an airport)? Does your company have a heavy reliance on E-commerce? Does it utilize SAP software, for shipment, engineering data or batch-tracking functions (a pharmaceutical company is a good example)? Is your organization a supplier a public utility, such as telephone service, electricity, water or natural gas? Does your firm perform electronic trading of commodities (thereby falling under the authority of the Sarbanes-Oxley Act)? If the answer to any of these questions is "yes," then your focus should be (at a minimum) Tier-III, or (preferably) Tier-IV.

We'll get into the details, costs and some examples of the different tier ratings, later. For now, it's enough to be aware of what they are, and how they might be applied to your business continuity needs.

2 MANAGEMENT OF THE DATA CENTER

Data centers are inherently difficult things to design, build and operate properly. From an overall perspective, it is as demanding to put together a management system for such facilities, as for building and reliably operating a jumbo jet. It could even be argued that the demands are even more stringent, since a jet airliner can be taken out of service whenever needed, whereas a data center is expected to run reliably and safely for years, without any major work being done. When major work is required, the data center is typically expected to maintain production operations, even when the work being done involve significant redesigns, upgrades and/or repairs. The customer simply assumes that it will *always* remain available, and they typically will not allow any planned interruption in their services.

To meet such demands requires four key facets of data center facilities operations:

- PEOPLE
- POLICIES, PROCESSES AND PROCEDURES
- SYSTEMS
- PLANNING AND COMMUNICATIONS

These facets are at two distinct levels: tactical and strategic. We'll call tactical anything within a 1-year horizon, and the strategic level >1 year in the future.

All of these pieces must be in place and functioning very well, if a high-reliability data center is to stay operational, and do it in a cost-effective manner.

This book will go in depth into each facet of what is required in the high-availability data center, first as a high-level perspective, and then we'll break down the individual components to deliver the desired result.

PEOPLE

The first part of this book will deal with your people; the engineering staff.

We'll look first at staffing considerations for a data center. When it comes to what level of staffing presence is required, there are a variety of industry recommendations regarding staffing levels for certain Tier levels, which generally can be followed. As you your requirements for operational uptime increase, you will have to supplement your manpower correspondingly. But what happens if you have sufficient capital money for your facility, but not enough in the operational budget for long-term manpower requirements? You can balance this equation, but it's a bit of work. We'll get into that.

Next, we'll examine alternatives to use when your staffing levels prohibit regular equipment rounds, in a less than ideal facility. We'll get into what kind of people you should look for (qualifications and experience, attitudes) in prospective engineering team members, and we'll examine what a training program should have (I will include an example program which you may modify for your needs), and we'll get into how you can best recruit them to your organization.

We'll get into the organization itself; what you should have for an organizational chart, critical facility descriptions, a roles and responsibilities matrix, typical designated key positions and how to successfully integrate this group show that it can affect am report up through the organizational structure.

A crucial factor is the weighing of training levels versus procedure dependence.

From a strategic perspective, how should the team work? What outside players need to provide regular input to the engineering team, so they can respond to shifting strategic needs? This will also address budgetary issues (OPEX and CAPEX), unplanned maintenance and so on.

POLICIES, PROCESSES AND PROCEDURES

Management consultants already know that most managers use the words policies, processes and procedures, interchangeably. The reality is that these words have *very* specific meanings, and are very *different* from each other. We must establish with clear specificity what these words mean, and how they apply to the management of critical facilities.

Policies are the general guidelines that we use to drive our processes and procedures. These policies will include things like access protocols, how maintenance programs will be administered and managed, housekeeping policies, life-cycle planning and the expectations of vendors in their support of the facility.

Processes are the management level view of what tasks need to be performed, and *when* they are to be performed. The processes are used for the scheduling and tracking of maintenance items, training, drills and other functions that can be planned for. Processes can also be put in place on how to respond to unplanned events such as equipment failures, service outages and other issues that might otherwise impact IT operations.

Finally, procedures which are the detailed steps on how an activity is to be performed within the process. In mission-critical facilities, the procedures must be very detailed, because the nature of the facility does not allow for mistakes were deviations from a proven script. We'll get into standard operating procedures- what they really are, what they should include, and how to verify that what you have (or are in the process of developing) will be adequate. We'll go into emergency operating procedures (EOPs), what they should include, how to train your staff on them, and how to get meaningful practice on them, so they'll be ready to act when the time comes. Other procedures will include how to create equipment logs and log-sheets for engineers to use when they're performing their rounds (equipment checks).

SYSTEMS

The systems portion will go into basic electric plant designs, emergency generators and uninterruptible power supplies (UPS). While there is a lot to the systems portion, we'll boil it down to the basics, so that a non-engineer (like myself) can understand what they mean, and weigh the different options. We'll examine the "sizzle" sold by capital-equipment salesmen and compare that to reality, and we'll even do a cost-benefit analysis of different options available for each approach to the infrastructure components, so that you can make truly educated decisions on what will work best for your application. We'll also do the same with mechanical systems- cooling and ventilation. Other aspects of systems include things such as where to build a data center, physical security issues and solutions, utility availability factors (water, power and fuel) and even natural-risk factors like seismic or flood risks.

PLANNING AND COMMUNICATIONS

This is the *most important facet* of managing mission-critical facilities;

without truly addressing this piece of the puzzle, you'll face far higher costs and lower performance of your facilities. Even if you have a "design-build" contractor come in a build you a state-of-art facility, if there is inadequate strategic planning and effective communication, you are doomed to suffer from less-than-optimal results.

The planning problems start with the simplest of questions. For example, a typical mistake of IT departments is they may want a mission-critical facility, but they aren't sure how to specify their needs in "engineer-speak." IT managers tend to look at time lost service, data recovery and expected time to recovery. Service Level Agreements (SLAs) with various IT-department customers aren't written in a format that building engineers can easily understand and translate into a physical reality. We'll examine this carefully, and hopefully clear up some myths about what "tier" a data center really is, what it means- and what it typically costs (capital expense dollars). We'll also examine the operational expenses of a data center, and how to realistically budget OPEX dollars.

Facilities teams usually maintain exceptional levels of communication and cooperation, within data center environments. Yet, the majority of data center owners and operators are aware of the fact that they're constantly running out of power or cooling capacity, or even floor-space. Why is this so? Data center facility managers typically have "power monitoring" and space-planning models in place, so why don't they work?

This is the secret piece of the puzzle that most data center operators haven't been able to solve, much less adequately address. The symptoms of inadequate planning and communications have become manifest in the data center (facilities) world, and the accompanying cost is *billions* of dollars per year in the IT industry... My research has documented the symptoms, and determined the ***actual*** cause. The solution will be presented with quantifiable results both in terms of reliability of the facility and costs, as well as a hard look at whether the IT demands meet the often-promised offerings of data center designers and co-location providers.

SUMMARY

A former manager once told me that having an MBA qualifies a person to manage any department, group, whatever. I got into trouble when I mentioned that having an MBA would not allow me to effectively run a data center, any more than it would qualify me to run the emergency-room at the local hospital. If you came in with a broken arm, I doubt you'd be impressed by my MBA or ability to read your statement of cash flows.

Mission-critical facilities such as data centers <u>must</u> be managed by technically strong leaders who also have above-average knowledge of business (such as an MBA). Like running the hospital ER or an airplane factory, you have to know exactly what jobs are happening in your facilities, why they matter, what the risks are and how to mitigate them. And, you must understand the core business enough so that you can effectively communicate with the IT clients, C-suite executives and project managers. The last characteristic you must look for in a critical-facilities manager, is bullet-proof integrity. Because you're going to be putting a lot of trust on this guy to keep you out of trouble with regard to tactical operations. [Indeed, it's more likely that you'll find an engineering manager/chief and have to educate him on the business aspect. Experienced business managers, interestingly, rarely make good managers for mission-critical facilities, because they don't have sufficient technical knowledge to know when the latest flavor of hype from a salesman (or consultant) is more "sizzle" than substance.] Finding this leader will be extremely difficult, and the candidate search won't be cheap. But you're going to need a leader to rely on, to protect your interests, in an area that's outside your area of expertise.

So the purpose of this book is not to make you an expert on the intricacies of data centers; it is assumed that you don't know the nuances of UPS designs (delta conversion versus double conversion), generator ratings, DX versus chill-water cooling systems and airflow calculations. We'll get to all of those things- and more- but at a high level so you know what to look for, what to avoid (and why), costs and other considerations for short versus long-term planning (tactical versus strategic). To finish it up, there will be addenda showing a comprehensive training program that can be copied and used in your facility, as well as a research paper demonstrating a key gap in IT management industry standards, and quantify the results of plugging that gap.

While I have been in the business of mission-critical facilities/engineering for >25 years, I'm not a degreed engineer. But as a friend once told me, "you don't have to know all the right answers, to be a good engineer; you have to know *where to look* for the right answers." And as I've learned the technical aspects of how the systems work, I've learned there are some very simple and direct questions that will tell you 95% of what you need to know. As we discuss the various aspects —people, procedures, systems and planning/communications- I'll give you those 'make-or-break' questions, their basis and what you should be looking for.

As in the real world, there is no written test. The test is knowing where to find the answers, when you need them. And my goal- my one TRUE goal, in writing this book- is to give you a single source where you can find those answers easily, in a simple yet meaningful format.

3 PEOPLE

CRITICAL FACILITIES STAFFING

The typical model associated with how many people are needed to staff a critical facility, is shown in the following example:

When you're operating a 24x7x365 facility, thus, (24 x 365) = 8,760 hours are required for each FTE.

Assume	Annual Hours	Not Available for Work	Comments
Work Year	2080		
Holidays		72	~9 days/year
Vacation/PTO		120	~2 Weeks' Vacation +PTO
Lunch + Breaks		260	30 minute lunch + 2 15-minute breaks
Training		96	Assumes 2 hours/week for 48 weeks
Total	2080	548	1,532 available for work annually/FTE

So 8760 hours/1532 man-hours per FTE results in:

FTE's required for desired 24x7 Coverage	Full Time Employees	Round up to
1	5.71	6
2	11.4	12
3	17.15	18

Notice that even though we go over 11 employees and 17 employees by a small amount (for 2 and 3 FTE's respectively), we still round up to allow for the off-going shift to communicate conditions or work being done to the oncoming shift; turnover. Current industry best-practices require a minimum of two FTE's on site at all times. There should be at least one, to monitor a facility during off-hours, but a single FTE on the second or third shift should not perform maintenance functions, as there will be nobody available to assist if the FTE is injured in the performance of such maintenance.

Operational expectations of facility:

Tier III data centers require a minimum of 1 FTE at all times.

Tier IV data centers require a minimum of 2 FTEs at all times.

Keep in mind, this model is useful only to calculate the *absolute minimum* requirements for FTEs. There will be additional considerations, such as how long it takes to do things like preventative maintenance, corrective maintenance (repairs and upgrades), building monitoring/data analysis, facility rounds, and the installation or removal of IT equipment, etc.

Overtime considerations

Over a long period of time ~10% leads to staff fatigue and potential health issues

Over long period of time, >10% causes risks due to human error

Preventative Maintenance (PM) is a key component of uninterrupted uptime. The man-hours required to fulfill 100% completion of PMs can be derived using both historical data for the different maintenance tasks of each piece of infrastructure, from the simple replacement of air filters to breaker remove-and-replace tasks. OEM guidelines can also supplement the historical data. The man-hours for PMs must also be broken out by trade (mechanical vs. electrical vs. fire/safety/security specialties)

- How many PMs are being performed?
- How many of the PMs will be contracted out to external vendors vs. in-house?
- How many vendor escort-hours will be needed to complete vendor PMs?
- Corrective Maintenance is equally crucial to uptime. Historical data and the current physical condition of the existing facility, must be factored into the equation to determine the likelihood of component failure.
- What is the condition of the facility?
- How diligently were PMs done in the past?

- What are current loads vs. original design maximum loads?
- Is the facility "occupied" with offices and other staff? If there are tenants in the facility who also require the assistance of the critical facilities team, the needed man-hours to fulfill those tenant work orders will need to be included in the overall total.
- IT Project support can also come into the mix; such tasks can involve the receipt, un-packaging and installation of IT equipment, as well as power-whip installations, floor-tile modifications and assisting with network cabling. This time will necessarily depend on the needs of the IT client(s).
- Work order count?
- Expected IT equipment refresh-rates?
- What is the generally expected level of support that IT clients expect when installing IT equipment?
- Vendor Management or escort responsibilities: vendors rarely are allowed to access critical areas without a security escort or someone to assist them with technical needs and/or information.
- Building rounds/Monitoring/Trend Analysis:
- How many physical reading must be taken in the facility?
- How many reading can be taken via Building Automation/Management System (BAS)?
- What are the failure modes of the pieces of equipment and alarms given?
- What is the physical size and layout of the facility?
- Periodicity with which physical inspections are performed according to criticality:
- Highly critical (little redundancy or remote monitoring) = 3+ times per day
- Medium criticality (some redundancy or remote monitoring = 1+ per day
- Low criticality (highly redundant or nonessential) = daily and during PMs
- Administrative requirements- how much paperwork needs to be filled out prior to and upon completion of repair/maintenance/corrective actions? How much time is required to get permission from IT stake-holders? How much flexibility is allowed in the *timing* of those actions?

While these aspects seem relatively minor, the effects to labor utilization to perform tasks are cumulative, and add up very fast. It's not unusual in a production data center to see double- or even triple- the minimum required in the facility, just due to the customer communication needs, physical size of the facility and materiel condition.

All of the above calculations also ignore the Critical Facilities Manager and other support roles for the team, such as a planner/scheduler, technical supervisors and other junior people who may (or may not) be qualified to perform all of the maintenance tasks that a critical facility demands.

CRITICAL FACILITIES STAFF ORGANIZATION

The Chief Engineer: He is the subject matter expert on all systems within his facility. He is responsible for everything that affects the facility, including electrical and mechanical systems, fire and life-safety systems, structural and building management systems. The Chief Engineer is also responsible for trend analysis of existing infrastructure, monitoring and managing electrical loads and providing proper oversight, supervision and training of all facilities staff and their work, both planned and unplanned.

The Assistant Chief Engineer, if applicable, is responsible for the daily operations and management of the critical facilities staff, especially staffing, training, and work performance. The Assistant Chief is usually also responsible for managing corrective and preventative maintenance operations, and working with both vendors and clients to insure that work is done properly and communicated to all stake-holders.

The planner/scheduler, if applicable, supports the administration of the facility, by tracking and managing things such as work-orders, computerized maintenance management system (MMS), man-hours and payroll, vendor invoices, warranty information, and equipment calibration.

The shift lead engineer is the senior qualified engineer of the shift, responsible for the performance of the technicians on his shift.

This model closely follows organized labor format as it applies to building engineering teams in populated facilities, such as office-building, high-rises, etc. The duties and responsibilities- as well as mindset- are rather similar. This in turn leads to the question of where to recruit critical facilities engineering teams.

RECRUITING CRITICAL FACILITES PERSONNEL

An underlying- and somewhat unpleasant- truth is that there are distinct differences in quality between "typical" building or "stationary" engineers, and those required for mission-critical facilities. The latter group, to be effective, have a near-obsession to performing to 'perfect' standards, and are loathe to accept less than the very best results and performance. This desire for high quality results tends to drive the best critical facilities engineers to learn more, and become experts in their chosen area of concentration. Thus, the labor rate of such staff tends to be significantly more expensive, as much as a 30% premium over 'normal' stationary engineers.

Typical places to recruit critical facilities engineers are through professional associations (linkedin.com) and local unions, as well as local trade-schools. Local unions offer experienced electricians and mechanical experts, but as of this writing there is no union-recognized certification process for mission-critical facilities, thus no assurance that the potential hire will be able to meet top-shelf expectations. Trade schools are an excellent option, to find people with the basic technical training, but without a lot of the "bad habits" that more experienced technicians may exhibit. Also, tech-school grads are typically hungry to learn more; they generally recognize that the tech school they graduated gave them the basics, and they'll have to work and learn more, through their own efforts and on-the-job training.

An excellent place to find well-trained engineers is through the military job boards, to "catch" trained and experienced mechanics and electricians as they are released from military service. As a military veteran myself of the US Navy nuclear power program (submarines), I can say with confidence that those who served in this program are highly trained, strive for the absolute best in results, and are sufficiently experienced to handle most emergencies with aplomb.

Finally, critical facilities engineers can be found through recruiters who specialize in this particular area, such as Peter Kazella and Associates (www.pkaza.com).

TRAINING

Training is not as easy as it seems. First of all, you need the staff to have at least the basics: working knowledge of tools, working knowledge of electricity and/or mechanical systems, safety protocols and applicable regulatory guidelines that they'll have to follow within the data center.

To find out where your staff is really at, you can utilize skill and technical testing, and reading/writing assessments. While it might sound unfair to require the latter, the reality is that most technical information including procedures and manuals, as well as schematics and piping diagrams, are written in English. The inability to competently read technical information could result in damage to equipment and severe injury or death of personnel. This is not something that should be taken lightly. To achieve success, staff members needs to understand the minimum knowledge that is required to safely operate and maintain the equipment and systems that support the data center and the entire facility. Note that this says the minimum knowledge required. Knowing or being able to do more is encouraged, but a staff members must learn what is needed to keep their installation running and what has to be done if something goes wrong. The objective is to minimize, or eliminate, downtime that negatively impacts company's ability to conduct business.

Following this initial assessment, staff need to be fully informed on regulations that the facility follows under, such as smog/pollution controls and protocols, fuel and hazardous materials storage and disposal, spill-containment procedures and equipment and basic medical care (especially for electrical shock). Sufficient study materials need to be provided to convey the general knowledge that all staff members need to know. Completion of these study materials should include a go/no-go test, and a signature acknowledging the expectation of compliance with all regulations.

When initial general training is completed, the Chief Engineer should assign an appropriate study plan for the staff member to complete, with established time-lines and fully communicated expectations. The study plan should be given to the staff member after the initial evaluation is complete, and after subsequent evaluations. It should detail from initial learning to final certification, and include both the knowledge and performance levels expected for the position. Time must be allotted for the staff member to study, meet with qualified staff members for training/testing/performance of tasks, and receive signatures from qualified staff when the tasks are complete. (This is referred to as a "Qualification Card Program," or "Qual-Card" for short.) The qual-card will have assignments of systems to learn, expectations of knowledge levels, standard operating procedures and emergency procedures. To enhance learning, signatures will be required not only for demonstrating sufficient system knowledge, but also performing tasks (such as completing the monthly PMs of air handlers) or operating switchgear, and simulations physically showing how to perform emergency procedures (such as manually starting a diesel generator).

Provided training material must include basic, technical and architectural drawings as needed, electrical and mechanical system layouts, and access to CAD drawings, for in-depth study and research.

New staff members will be assigned to work with qualified critical facilities engineers for 4-week period, until they rotate through all the qualified engineers, so that they get a firm grasp on all of the tasks that are performed in a data center by each critical facilities engineer. As well, this approach provides wide-ranging on-the-job training (OJT).

Periodic reviews should be performed between the Chief Engineer, the new staff member, and the assigned critical facilities engineer that the new staff member is assigned to work with. As well, the reviews should include testing of learned systems, "what-if" questions, true/false quizzes or whatever else is deemed effective. Feedback to the new staff member needs to be constructive, to help the staff member succeed, instead of criticizing and setting that person up to fail.

Occasionally, a new staff member will be unable to sufficiently grasp system knowledge or perform to the level of expectations of the rest of the critical facilities staff. In such a case, a poor match of capabilities versus expectations can cause frustration, as well as jeopardize the ongoing operations of the data center. Thus, evaluations need to be honestly performed, and the Chief Engineer prepared to reassign the staff member to other functions which the member may be more suited to.

As a final note, the newly hired staff member needs to have sufficient time allotted for them to meet the performance expectations of the qual-card program.

A detailed qual-card program is provided at the end of this book (Appendix IV). If needed, the document can be copied and utilized, as I make no copyright claims to this (it is public-domain). If your facility lacks a robust training program, this provides a very powerful beginning.

ATTITUDES OF CRITICAL FACILITIES STAFF

The employment of top-notch critical facilities staff results in some interesting side-effects, that the IT manager or executive should be aware of. Critical facilities engineers- and especially Chief Engineers- become very good at their roles, because they take an interest in their work, have a superior work-ethic, and become truculent when forced to work with either substandard or equipment or results. With decades of experience, they tend to behave like prima donnas, generally regarding the facility they maintain as their exclusive domain. I've seen a few exceptions to this fact, but it's largely true.

Thus, their behavior can appear gruff and crude, even arrogant. One chief I managed, I worked with extensively, to keep his comments, complaints and insults (to customers he didn't like) under cover; he instead used me as his sounding-board for various issues. While it was not a particularly pleasing situation, he gave me plenty of information about behind-the-scenes activities with various clients, I was able to rework his publicly perceived personality so the customers were no longer offended by his previously argumentative attitude. Most importantly, we were able team up as colleagues to gather budgetary information and risk assessments, and get sufficient funding to solve some of the long-deferred maintenance that was bothering him. As I told him on more than one occasion, he was a serious thorn in my backside, and was driving me crazy... but he was the most valuable man on the team, because he pushed hard to make sure that the data center uptime was guaranteed.

This last bit is a perfect example of the prima donna attitude in action. IF you view those people- who might at first appear arrogant and insufferable- as an asset and actively listen to them, excellent results can be obtained. It will challenge your levels of patience, but the rewards can be well worth it.

OTHER CONSIDERATIONS

If you have a critical facilities portfolio, where you are dealing with a few large facilities and the rest of them are smaller and cannot justify the presence of FTEs, you may have to rely upon the services of a contracted firm (or several) to handle emergency calls and/or maintenance issues.

This can be a viable option in certain cases, provided the contractor work is supervised and signed-off by a competent manager or chief engineer, who MUST verify that the work is not only done properly, but is a legitimate expense for the maintenance budget. If the work is not properly supervised, contractors usually figure out that they can "pad" their expenses without it being detected, and expenses will stay the same- if not increase (!)- while actual results decrease. Sadly, this is human nature; people will tend to take advantage of a situation when given the opportunity, so diligent supervision for outsourced services is a must.

As an example, on one particular project I was involved with, the vendor charged the project budget for overtime work on a Sunday (doubled labor rate); it was later discovered those persons who were supposed to be "working" were actually at home watching a football game! To say the least, they didn't get paid for their "overtime!"

4 POLICIES, PROCESSES AND PROCEDURES

There are some basic policies which MUST be carefully considered for the mission-critical facility like a data center.

MAINTENANCE TASKS- MANAGEMENT

Critical facilities technicians are tasked with three functions: monitoring the critical environments and their support systems, performing and managing maintenance activities, and performing emergency actions to stabilize the support systems in an emergency, until help can arrive. Maintenance management takes up- by far- the majority of man-hours in a critical facility.

To effectively manage maintenance activities requires a program to record, track and communicate the daily operations, as well as plan for future events. A Computerized Maintenance Management Software (CMMS) program allows for tracking the percentage of planned operations completed on schedule, as well as "child work orders" that are developed from existing maintenance operations. Key performance indicators (KPIs can be developed to track things such as percentage of downtime due to malfunction/broken machinery, and compare total costs of ownership against the lifecycle of the machine. CMMS has become the standard tool for managing these operations, and can manage literally tens of thousands of pieces of equipment, tracking data and performance with ease.

An effective CMMS will have the scheduled PMs associated with a piece of equipment, such from daily to annual tasks, but will also be able to record and maintain a history of all work associated with the equipment, maintain an inventory of maintenance parts for the associated gear, all other resources required to perform tasks on the equipment, and be able to effectively extrapolate data into meaningful metrics that manager can use to evaluate performance of the equipment and the effectiveness of the maintenance/repair personnel performing such tasks.

A CMMS should include methods to allow an individual to make a service request or trouble-call, which then becomes part of the maintenance management schedule for control and resolution. As well, most CMMS systems have a work-order tracking system to communicate out the status of the repair request, estimated time to completion (or when completed), and associated data to complete the work order such as parts and materials, labor, tools, costs and so on.

A CMMS should also have the capability for a planner/scheduler to schedule work in the future, determined by operational requirements or constraints, available man-hours, parts availability and even the technical skill-set of particular shift personnel.

Finally, a robust CMMS should have the ability to quickly and efficiently schedule preventative maintenance (PM) operations base on varying periodicities, such as daily, weekly, monthly, quarterly and annual PM schedules. These are often based on manufacturers' recommendations and the experience of the management staff. While the PM operations may be time-based, they may also be based on the amount of hours a piece of equipment runs, or whether indicators of that equipment point to increased frequency. The PM function should automatically generate the time-based PMs, allow the technician to download pre-verified SOPs to perform the task (technician must verify the SOP is signed as verified technically accurate) as well as the actual PM card to have a step-by-step guide to all actions to be performed.

Equipment manufacturers will always list recommended maintenance practices for the equipment they are selling. As an example, Caterpillar has the following recommendations for a generator that is in operation:

Daily	Every Week
Control Panel - Inspect/Test	Automatic Start/Stop - Inspect
Cooling System Coolant Level - Check	Battery Charger – Check
Engine Air Cleaner Service Indicator - Inspect	Electrical Connections – Check
Engine Air Precleaner - Clean	Generator - Inspect
Engine Oil Level - Check	Generator Lead - Check
Fuel Tank Water and Sediment - Drain	Space Heater - Test
Generator Bearing Temperature -	Stator Winding Temperature -

Test/Record   Test
    Generator Load - Check   Voltage and Frequency - Check
    Walk-Around Inspection

As well, they also list the following recommendations (partial list):

Every 250 Service Hours
Battery Electrolyte Level – Check
Belts - Inspect/Adjust/Replace
Cooling System Supplemental
Coolant Additive
(SCA) - Test/Add
Engine Oil Sample – Obtain
Fan Drive Bearing – Lubricate
Hoses and Clamps
Inspect/Replace

Notice that the list of maintenance tasks on the left are as a function of the number of hours the generator has been running, not necessarily as a function of calendar time. This makes sense because oil and grease doesn't degrade much (if at all) -with time, but it *does* over time, as the engine is running.

MAINTENANCE TASKS- PREVENTATIVE MAINTENANCE

Preventative Maintenance (PM) is a methodology of maintaining a piece of equipment in a proactive manner, by utilizing manufacturers' recommendations, historical trends and physical monitoring of the equipment in question. That is, it's a systematic approach to avoid problems before the equipment fails. There are two basic approaches to PMs; condition-based PM, which tracks the life of the equipment and condition, to make statistical models and predict when (and why) failure may occur, or scheduled maintenance, exactly like changing the oil in your car every 3,000 miles.

PM generally consists of periodically replacing consumable parts, system adjustments and equipment calibration, cleaning/replacement of filters, belt adjustments for fans, lubrication of bearings and/or the updating of software and firmware.

It should be noted that PM is not the same as Corrective Maintenance (CM), which is fixing something after it has failed.

PM is especially important in the critical facilities world, because the organized maintenance- which can be done at a time of your choosing- is infinitely preferable to attempting to repair a piece of critical equipment in an emergency (unplanned, high risk) situation. Critical facilities departments almost always use the scheduled maintenance approach, as it is easier to work with clients using an established calendar, and the guess-work associated with statistical failure probabilities is eliminated.

The PM program must include manufacturer recommendations including allowable adjustment specifications, consumables parts numbers and maintenance periodicities for each level of maintenance, from your basic daily checks (done via log-sheets when the technician is performing their rounds), weekly, monthly and annual PM operations. The PM program must utilize *fully trained* personnel, who are familiar with the equipment, safety procedures and tools to be used, and communications protocols.

Below is an example of a paper PM card for an air handler in a critical facility:

A quick examination will reveal that while this PM card appears to list off the various important items (lubricate bearings, inspect belts and bolts, check for cleanliness and any burned out indicator bulbs), there are a lot of key pieces missing:

- Safety equipment
- How to secure the piece of equipment so that it cannot be restarted (secure all energy sources), in accordance with OSHA Lockout-

Tagout instructions

- Equipment or tools needed for the job
- Specifications for adjusting the belt tension
- Type of lubrication to use with the bearings
- Communication protocols, from commencing the work, to completion.

This card, to be fair, is from a data center where the management refused to implement a computerized maintenance management system, so the Chief Engineer had to develop and implement the system on his own, without outside assistance. While rudimentary, it was still very useful for tracking maintenance within the facility, though some aspects like historical trending were nearly impossible to achieve, using this approach.

MOST maintenance management systems are of similar quality to the example above; minimal information for the technician to utilize, so he must learn through trial-and-error what tools will be required, what the machine "likes" (as opposed to catastrophic failure because the wrong type of grease was injected into the bearings) and how to safely de-energize the equipment to work on it. The PM systems from outsourced firms who specialize in building management also reflect this lack of meaningful detail, as well.

A proper PM card for a piece of equipment, MUST include the bulleted items above. As well, if the facility is very large, there should even be a floor-map so that the technician can quickly locate the piece of equipment to be worked on. In this way, as soon as a technician has proven technical competence in performing the applicable maintenance on a single piece of equipment (such as an air handler), you can then hand him the PM cards for all the *other* air handlers in the site, and he'll literally have all the information he needs, to complete every other handler of similar design in the facility.

When a schedule preventative maintenance action is to be performed, a search should be done using the CMMS system to find any associated work orders for deficiencies or problems which can be corrected at the same time the PM is being performed. It is a serious waste of man-hours and effort, to isolate a piece of machinery for scheduled work, and then forget to address other (perhaps minor) problems that might just as easily have been corrected at the same time. And while this sounds a bit absurd, in large facilities it is a common problem.

PM Results

When a PM action is performed, the expected result is most often that the operation is successfully completed, with no outstanding issues noted. This is, after all, the goal of the PM program. But sometimes a new or potential issue is discovered. In the air handler example above, during the PM, the technician notes that the tensioner holding the belt nice and snug for the air handler has a bolt coming loose. He tightens the bolt to remove the slack in the belt, and then discovers the lock-nut (which keeps the tensioner bolt from loosening up to vibration) has rounded edges, so that proper operation of the locking-nut is not able to be applied. In such an example, the technician would normally be expected to affect repairs immediately. In this example, the repair might take an additional 10-15 minutes. But what happens if the replacement part isn't available? Then it is up to the Chief Engineer to decide if the air handler should be taken out of service, if it can run without the lock-nut until a new one is obtained and repairs implemented, or simply note it to be repaired at the next maintenance window.

When unplanned conditions come up, the senior engineer should be consulted with immediately, so that he can determine whether the issue represents a risk to data center operations or not, and the best course of action to take to balance operational needs against risks of downtime.

If other issues are detected- either with the unit being inspected or other issues that are found by happenstance- this must be documented with a work order to be entered into the CMMS system being used.

MAINTENANCE TASKS: BUILDING ROUNDS AND EQUIPMENT LOGS

To properly monitor the ongoing conditions of a critical facility, a trained technician needs to periodically tour the facility, performing an inspection and logging equipment parameters.
1. The inspection is to visually check for signs of trouble or changing conditions, using four of the five senses:
    a. Smell, to detect
        i. Smoke
        ii. Leakage of liquids, like oil from a gear-box
        iii. Noxious fumes indicating chemical spills
    b. Sight, to detect
        i. Spilled liquids
        ii. Alarm lights
        iii. Other unusual circumstances
    c. Hearing, to detect unusual noises
    d. Touch, to detect unexpected heating/cooling problems with equipment
2. The logging of equipment parameters is to insure that technicians

are
> a. directly checking the indications of equipment, as well as
> b. building a historical record for trend analysis.

The time required to perform building rounds will depend on several variables.

The first variable is the history of the facility, and its physical condition. If the facility is in poor materiel condition, more monitoring will be required so that problems can be caught as quickly as possible.

The next variable is the actual design of the facility, including its physical size and how easy it is to access equipment for inspection. It is common sense that the larger a facility is, the more complicated the design and/or the more difficult it is to access the equipment to be checked, the more time will be expended to overcome these obstacles.

Building Management Systems (BMS) or Building Automation Systems (BAS) are designed to remotely monitor and/or control building systems. This can reduce the amount of time required for physical building rounds, but does not *guarantee* it; a technician using his physical senses can easily detect developing problems that may not manifest themselves in a BAS/BMS for hours or days. Thus, a BAS/BMS is <u>not</u> an effective substitute for physical inspections of equipment on a periodic basis.

The type of failure modes that a piece of equipment or system can suffer, also affect the time for inspection. A simple valve can have only a few ways to fail: leakage through the valve stem or flanges, failure to open and failure to close. Thus, a valve is easy to inspect. Inspecting an air handler on the IT floor can require more time, and the overall electric plant even more-so.

The amount of time can also be affected by regulatory requirements; local government regulations typically require the periodic inspection of diesel fuel tanks, in that they are not leaking and that they're level indications remain unchanged.

The more critical the equipment is to the facility, the more care should be taken to monitor its health. It's logical to devote more time to the UPS and cooling systems that are online, than a small auxiliary pump which is not in operation.

The physical inspection and logging of equipment parameters is very useful to verify that remote monitoring systems BAS/BMS are performing as expected, as well as to record data for trend analysis over months and even years. The data itself becomes a historical record that can be used as a tool for establishing future facility needs.

Effective log sheets must record data that is relevant and meaningful, without recording extraneous data. The data that is recorded must be specific, with an expected range (high and low) that the parameter must stay in. Equipment logs need to take into account the recommendations of the manufacturers. That is, they should reflect the appropriate design parameters of the equipment. The log sheets must include very specific instructions on how and when to take the readings.

Finally, the Chief Engineer <u>must</u> review the logs in a consistent and ongoing basis; parameters that are not within specification must be investigated according to the level of criticality of the piece of equipment, and overall conditions within the plant at the time of the condition. An established protocol must be in place, to properly communicate unusual conditions.

At one facility I was working with, the logs had been taken diligently for years, recording trends where the loading of certain electrical components went beyond plant design limitations. This slow trend created limitations in redundancy (and therefore reliability), reducing the operational readiness of the entire facility. The assistant chief engineer had reviewed the logs and was aware of the situation, but never connected the dots together, to understand the risks that were becoming more and more severe. This demonstrates why the logs need to be consistently reviewed, and compared to not only expectations, but ultimate design parameters.

PROCEDURES

This section lightly touches on Standard Operating Procedures (SOPs), Methods of Procedure (MOPs) and Emergency Operating Procedures (EOPs)

SOPs

SOPs are *incredibly* important to reliable and sustainable operation of the mission-critical facility. And while this seems to be self-evident, you might be surprised at how many data centers don't have full SOPs. Therefore, we'll take some time to go into what an SOP is, how it's developed and implemented, and how it's performed on a *continuous* basis.

An SOP is the standard operating procedure to perform typical tasks within the data center environment, whether that task is starting or stopping a UPS, rotating pumps or chillers, or performing maintenance tasks. SOPs should cover all conceivable engineering events which are normally expected.

What is included in the SOP?

The SOP should begin with the task being performed, periodicity and required skill-level of the operator. It should include a list of tools, test equipment and parts needed to complete the task. If it requires a high level of regimentation, the level of detail must increase correspondingly. It may include some or all of OEM recommendations and protocols within the SOP, as well as any additional documentation such as schematics, exploded drawings or specifications, as appropriate to assist the technician.

All components such as valves, switches or breakers, control buttons and indicators need to be properly marked on the equipment or system to be operated, and referred to in the SOP in a clear and concise manner. Each step should include only <u>one</u> action performed.

Each step in the procedure MUST be either checked or initialed by the person performing the work.

The SOP should be formatted in the following way:

1. Scope of Work
   a. Description/Name of work to be performed and on what piece of equipment
   b. Facility
   c. Purpose
   d. All attachments included with the SOP
2. Risks and Methods of Recovery
   a. Expected Impact(s)
   b. Risks and Methods of Recovery
3. Preconditions
   a. Staff Required
   b. Communications to potentially impacted clients
   c. Facility Status
   d. PPE
   e. Permits (such as fire or energized electrical work)
   f. Tools, equipment and parts required
4. Procedure (step-by-step)
   a. Step
   b. Expected results
   c. Data logging (as needed)
   d. Signature/initial line
5. Conclusion
   a. Communicate to affected parties
   b. Review work
   c. Document completion

The step-by-step process should be so detailed as to include things like operating a specific breaker, switch or valve in the exact sequence required and then stopping to initial that step *before* moving on to the next step. The step should include the expected result, such as when I press the "off" switch on a UPS, what indications would I expect to see on the display screen? Verify those indications, before moving the output breaker to the "open" position, by initialing each step of the procedure.

The SOPs can be largely duplicated amongst identical parts of system, with one exception; the valve/switch/breaker designations must be changed so that each SOP applies to that specific piece of equipment. For example, it's not unusual to have six condenser pumps in a chiller system, all of the same exact design. However, the valves that control flow to each pump will have different designations, so that must be reflected in the SOP for that *exact* pump.

Ideally, the SOP should be so detailed that you could give it to a junior technician who has never seen that piece of equipment before, and he'll be able to find it, perform the task without any clarification or additional help, and complete the SOP to its expected conclusion, all without issue.

The SOP should be verified as technically correct by a competent authority, and signed as verified correct. Periodic audits (typically 1 year) should be done to ensure the document has not been invalidated due to changes in equipment or processes. The signed copy should be uploaded in digital format so that when a technician goes to perform that SOP, such as when performing preventative maintenance, he can quickly verify that the SOP *is* verified accurate, and applicable to the task.

MOPs

MOPs, or Methods of Procedure, are one-off processes designed to accomplish a specific directive. An MOP is appropriate when your efforts to either isolate the unit or bring it back online may have an adverse effect to critical IT operations. For example, if you need to isolate a part of a system such as a UPS for maintenance or repair, the MOP should have steps to isolate that unit, which would typically be covered in an SOP (the MOP can refer to the appropriate steps of an SOP) and then allow latitude for troubleshooting, maintenance and/or repair. When complete, the MOP should include testing protocols to verify proper operation. After that portion of the MOP is complete, the procedure should conclude by having the steps to put the unit back into service (again, typically the same steps as an SOP would have).

While the MOP is typically a one-use-only type of procedure, like the SOP is must be verified as technically accurate and signed off by a competent authority, before being handed to the technician(s) to perform. The technician(s) must also verify that it is signed off as technically complete, before commencing work.

EOPs

Emergency Operating procedures (EOPs) are just as the name implies; what to do when a *foreseeable* event occurs that is outside normal operating conditions. Such an emergency could be due to smoke or fire, flooding or rupture of a city water-main, partial or complete loss of utility power, failure of systems within the data center or natural events such as tornados or hurricanes.

The purpose of EOPs is to therefore to either recover the facility from the emergency, *or* perform effective measures which will put the data center in a stable operational condition until an effective recovery plan can be determined.

Due to the unexpected nature of emergencies, EOPs must be concise, and formatted similarly to SOPs. The EOPs should include steps to take for a given situation, expected outcomes, PPE to wear, and communication protocols as the situation requires.

EOPs should be developed for all realistically likely situations that may jeopardize operations, personnel safety, infrastructure, the environment or client reputation. Single-point-of-failure analysis, as well as potential cascade-failure scenarios, will uncover other areas needing EOPs. Understand, however, that an EOP can only be written for a single emergency, *ceteris paribus*. EOPs cannot be written that can account for cascade-failures. If a single fault causes another component of marginal quality or effectiveness to then fail, and so on, your existing EOPs are no longer valid, and the engineering team must rely on their training and their wits, to safely and effectively stabilize the facility until recovery efforts can be implemented.

After development of SOPs, MOPs and EOPs, they must be implemented in such a manner that all engineering personnel can either perform or simulate the operations described in each procedure, to become familiar with the task(s) to be performed. EOPs should be conducted on a routine basis as standard training, to keep engineers and technicians comfortable with potential events and how to respond in a quick and efficient manner.

FOR ALL PROCEDURES, Each step in the procedure MUST be either checked or initialed by the person performing the work. Failure to properly validate completion of each step can have lethal consequences:

On August 31, 1988, Delta Air Lines flight 1141 crashed on take-off, killing 14 of the 108 passengers and crew on board, and injuring 76 others. The pilots were distracted in the middle of their pre-flight checklist, and when they returned to it they missed the step where they were to set flaps to maximize lift for low speed flight. The safety system for such an event, the Take-Off Warning System (TOWS), designed to alert the crew if the engines are throttled to take-off power without the flaps and slats being correctly set, was not operating correctly. The plane stalled on lift-off, crashed and broke apart 1000 feet past the runway, and burst into flames. Subsequent FAA requirements included physically marking the completion of each item on a check-list, as a visual reference to help pilots complete all items without any missed steps.

A precaution here: some managers and executives have the expectation that there will be a formalized procedure for every possible situation that can occur. Unfortunately, this is not only unrealistic, but a plethora of procedures causes its own unique set of problems; as the book of procedures grows in size, the *less* likely that technicians will fully read and understand all of the procedures. In effect, people reach a *saturation point* of procedures, beyond which they simply give up trying to memorize them all. Thus, having voluminous tomes of technical procedures gives management a false sense of security, while the critical facilities engineers will either be disinterested in those procedures, or afraid to work on the equipment due to fear of punishment for not following an unknown procedure.

Wrap-up of Procedures should include a full summary of what was done, recorded parameters (and make sure they are within expected ranges), any unusual conditions detected, and any reports which the equipment vendor might have recorded (battery voltages on a UPS string, for example).

5 SYSTEMS INTRODUCTION

As I mentioned at the beginning, you cannot effectively manage a mission-critical facility or portfolio without having technical expertise in the field, any more than a person not medically trained can run an emergency room at the local hospital. Such technical expertise comes from *decades* of work around critical facilities. Even then, well-seasoned critical facilities engineers rarely become qualified to be chiefs, because the transition from being a specialist (in electrical, mechanical or building automation systems) where you looked at individual trees, changes when you are now expected to watch the entire forest. The need to have deep understanding of *every tree* in the forest, is required, to manage the forest overall. It's not at all unusual to meet facilities engineers who have been doing the same type of work for 30 years and never became chief engineer, either because of trepidations at the thought of total responsibility (and personnel headaches) or because they simply were incapable of having a holistic view of their mission-critical facility.

The manager of the chiefs- a regional data center manager or director- must have a similarly holistic view of systems and a medium to high level of knowledge, to understand what the chiefs are telling him/her, and how to respond. As well, the director must be able to examine the information provided by the chiefs, analyze the data, and translate that into actionable information that the IT clients can utilize. This might be an emergency capital-repair project, it may be the enhancement of existing facilities to accommodate new IT power/cooling/space demands, or it might be as simple as trend-analysis to get a handle on how long the client has before they run out of power, given an established growth-rate.

This section- Systems- is specifically written to give you a high-level overview of what is inside a mission-critical facility like a data center. It is *not* all-inclusive, so have no illusions about that (or concerns that you'll be buried in minutia). It is base-line information that you'll need to understand what is in the facility, how it works, what the "good" and "bad" points are of different approaches to a particular problem. This will allow you to make better-informed decisions on what should done in your situation, with the input for your chief engineer(s). I have taken this particular approach, because there are literally hundreds of different designs, and dozens of industry equipment manufacturers, each with their own methods to solving certain issues. As a result, even a UPS expert can only specialize in a few designs or brands, because each brand requires an *insane* amount of in-depth knowledge. [Your chiefs will have a deep level of system knowledge with their own experience-based views, which you must leverage to obtain all needed information for effective decision-making, while weighing your company's tolerance for risk.]

I have included also included with each portion of systems, a summary of the good and bad of various approaches, as a "cheat sheet" to use when looking at something. Finally, with the description of each system is a list of basic "do's" and "don'ts" that would *seem* self-evident. However, I included them because there are a LOT of small details which can have a large cumulative effect on energy efficiency, cooling effectiveness, reliability, redundancy and so on. It's not to insult the reader or the chiefs that work for you; instead, they're checklists to make sure that all the simple (cheap) solutions can be utilized at the beginning, to give you the best bang for each dollar spent, and making sure that your facilities are as good as they can be, before considering spending money on high-dollar solutions or improvements.

The last thing to mention as an introduction to Systems, is that various vendors "sell the sizzle" when it comes to their wares. That's what they *do!* I'll include my own experiences where possible, adding a dose of reality to the picture. For now, it's enough to say that while the promises made by vendors may be valid in *controlled* conditions, the validity of such promises become questionable in real life.

6 SYSTEMS: SITE SELECTION AND PHYSICAL CONSTRUCTION

The first and most obvious system you must consider, is the physical building itself, and where it is located.

Suppose your business model justifies a new data center. You've run the financial calculations, looked at the justification for such a move, and are now seriously considering the construction of such a facility. This brings into play a new set of questions that can be daunting. Where to put it? How big? What kind of security will be required? How good are the local internet providers, and how much bandwidth is there?

Your internal IT personnel should be able to give you solid information on the last two answers, but the questions of where, when and how, are things that we can directly get into.

NATURAL FACTORS TO AVOID

When selecting a site the first rule is to avoid any areas which have a medium to high risk, for natural disasters. Some easy examples are to avoid building your facility in a flood-plain or an area where there is a history of earthquakes, hurricanes or tornados or extreme snow/ice.

Other risks to avoid are areas where there is a history of flooding, due to ice/snow thaw or heavy rains. Avoid at <u>all</u> costs building downstream of a dam or levee. (A civil engineer once divided levees into two groups: those that have failed, and those that *will* fail). Make sure the critical facility and all associated support equipment are above the 100-year flood plain.

Other things to avoid are mountains and steep hills, as they make security more difficult, and increase construction costs. Avoid locations where utility services cannot be fed from different power grids (part of infrastructure redundancy). The same applies to communications lines.

Another risk to avoid is any location where water is a scarce commodity (we'll delve into the reasons behind this in the chapter on Heating, Ventilation and Air Conditioning). Instead, look for a location- if possible-with a moderate-to-cold climate.

Avoid seismic zones.

The final risk to avoid is any location where known soil-contamination issues exist, or the soil is unable to be drilled for geothermal cooling (a highly efficient, "green" method of cooling).

MAN-MADE FACTORS TO AVOID

Airplanes sometimes crash, especially while landing or departing from airports. While it doesn't happen very often, it is important to minimize this risk. If you must have your critical facility in near proximity to an airport, at least make sure it is *not* in the direct flight-path to any of the runways. This is true whether for an international airport, or a small-aircraft facility. The best practice is be at least 5 miles from the typical flight path of any nearby airport or the airport itself.

The next man-made risk to avoid is being near high-traffic areas, such as highways and interstates. Better to be at least 1 mile away from such areas, so that an accident such as the crash of a truck carrying dangerous chemicals or explosives will not endanger your facility.

Avoid locations where the rapid acquisition of generator fuel is problematic. If your site is in a seismic zone, it becomes important to have access to the facility from the main roads while avoiding bridges and roads that may be damaged during an earthquake. Having nearby sources of reserve generator fuel is not helpful, if you can't physically deliver it to your site.

Avoid locations where you cannot establish a perimeter of at least 180 feet around the data center. In other words, make sure outside (street) traffic cannot approach to within 180 feet of your facility.

Avoid locations where ready access to semi-trucks is difficult or impossible, and avoid high-crime locations, to minimize risks of sabotage/terrorism/theft.

PHYSICAL SECURITY AND PROTECTION

In our personal lives, our own physical security begins with common-sense. In other words, we don't explore high-crime areas, don't flash a wad of money at the local mini-mart, and don't use an ATM in a dark alley. Simply put, don't *advertise* to people who may have unpleasant intentions or proclivities that you are a potential target.

Data centers- by their nature- are the Achilles' heel to the large corporation. And large corporations, almost always, have their detractors and self-avowed enemies. In such a group are those who would gain some measure of satisfaction by harming the profitability of the company, if not downright crippling.

The methods to harm the corporation include obvious attacks on networks and internet security and viruses. This threat allows a measure of anonymity to the attackers, with the trade-off that protective measures can be effectively utilized to minimize or eliminate the impacts of such attacks. The goals of these tactics, are usually espionage, data theft or sabotage.

There are people, however, who will be more daring, and attempt to harm the corporation by more direct (and dangerous) means- physically infiltrating the data center security or sabotaging infrastructure equipment.

Physical attacks to consider include arson, using a vehicle to ram a building or its support systems (such as outside generators or cooling towers), explosives or firearms.

And while our penchant for watching "shoot-em-up" movies makes us think of scenarios like explosive devices or other outlandish attacks, there are ways to physically cripple a data center, which are much more quiet, and are still effective... Indeed, the latest threat discussed in the U.S. Senate (during July 2009, prior to passage of the Cybersecurity Act of 2009), is the use of "Intentional Electro-magnetic Interference" (IEMI). The more common term for this, is "electro-magnetic pulse," or EMP.

EMP "bombs" can effectively be built using commonly-available materials from an electrical-supply outlet, and are nearly impossible to trace after use. Sound exotic? Try a price tag of ~$800. An EMP bomb can be fitted within a suitcase, and still retain sufficient power to kill all electronic devices within a 60-yard (180 feet) radius. Should a person decide to spend the money to obtain a surplus radar dish from a yacht or fishing-trawler (this component only costs about $500, and is available on Ebay), it becomes possible to construct a directed-energy weapon with an effective kill-range of up to 600 meters, and will easily fit in the back of a pickup truck.

Unless you want to build your data center out in a cow pasture, where a 600-meter perimeter is easily possible, the options for defense against such weapons involves internal shielding/grounding of IT equipment, or make sure the people who wish to harm your firm simply don't know *where* your data center is located!

In short, you need to practice the same methods of protection for your data center, as you practice for your personal protection; don't advertise.

DATA CENTER SECURITY BEST-PRACTICES, EXTERNAL

Your security measures should be based on concentric circles, starting with location and exterior appearance.

LOCATION: The data center should be isolated from any public-access roads, as much as possible.

PERIMETER: Maximize the perimeter to your building as it is the first layer of your data center security. Traffic approaching the building should be on an as-needed basis, and no storage of cargo vehicles should be allowed within the vicinity. Avoid the use of concrete barriers to shield the facility from vehicles, as this attracts attention. Steel bollards and heavy chains between them will accomplish the same task, with less visibility. In the example below, the shaded area represents your primary layer of physical security:

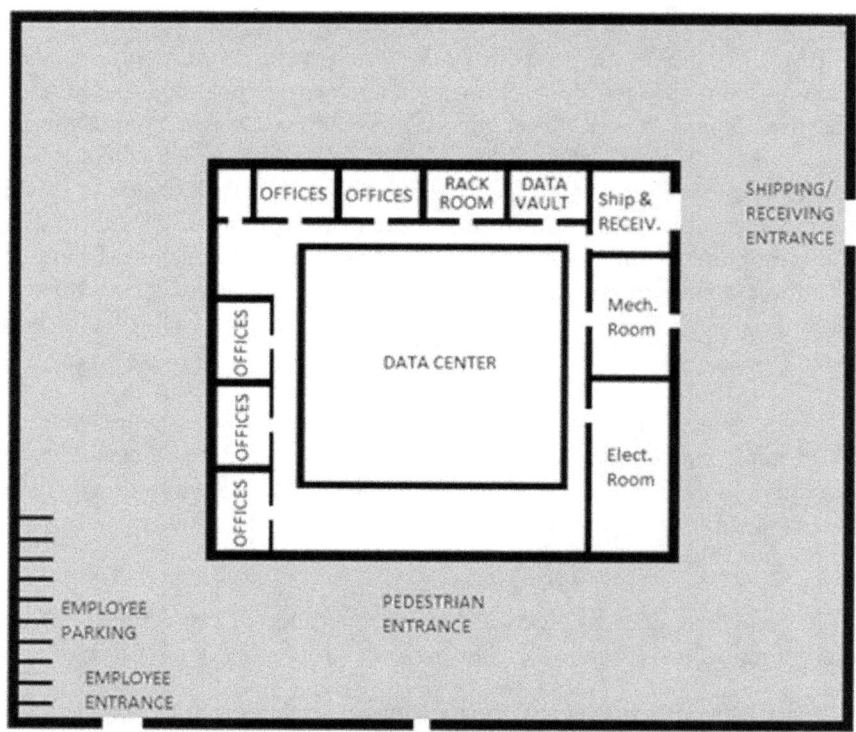

<u>EXTERIOR</u>:   The data center building should appear dull and uninteresting. It should be isolated from other facilities, with no external windows to provide weakened architectural locations for unauthorized entry, or provide direct access to IT or infrastructure spaces.   The importance of the facility, in effect, should be minimized by external appearances, to avoid unwanted attention by persons who might wish to harm the company. When you read the physical description of the top-of-the-line data centers which cater to prestige clients, the term you will see repeatedly, is "nondescript."   Ideally, an outsider should look at the building, and think it is a low-level storage area of some kind.

Avoid having bushes and trees next to the facility, and keep them at least 50 feet from the building walls. This will minimize the ability of a troublemaker to gain access to the building unobserved, and eliminate the risk of trees falling on your facility in the event of a large windstorm.

Keep areas close the building free of trash receptacles.

Minimize the amount of parking that is close to the facility. Remember, the goal is for the data center to be effectively invisible, to outsiders.

LAYERED ACCESS: Access within the outermost perimeter should be through a vehicle-proof gate via a card-reader or with security operation (after verification of need via delivery schedule and CCTV to verify that person is supposed to come in). This is your first line of defense, and very important to have in place. Do not have a gate that is left open, as it compromises the entire purpose of your security measures.

Building access should be limited to only those that need to be there. The delivery driver should have a security escort at the loading dock, and never be allowed further. The network operations center (NOC) operators should not have access to the production floors of the data center. Actual data center access should be strictly limited to only those people who must perform physical tasks within the data hall, such as rebooting servers, installing new power whips or maintaining infrastructure systems. Visits within the data hall by non-essential personnel (including IT managers, Network Operations Center (NOC) personnel and even IT customers) should be granted only on an as-needed basis.

Indeed, from the APC/Schneider whitepaper "Physical Security in Mission Critical Facilities,"

When data center security is mentioned, the first thing likely to come to mind is protection from sabotage, espionage, or data theft. While the need is obvious for protection against intruders and the intentional harm they could cause, the hazards from ordinary activity of personnel working in the data center present a greater day-to-day risk in most facilities.

People are essential to the operation of a data center, yet studies consistently show that people are directly responsible for 60% of data center downtime through accidents and mistakes — improper procedures, mislabeled equipment, things dropped or spilled, mistyped commands, and other unforeseen mishaps large and small. With human error an unavoidable consequence of human presence, minimizing and controlling personnel access to facilities is a critical element of risk management even when concern about malicious activity is slight.

While legacy-thinking says that IT people have to be in the same building as their equipment, this is simply not true, anymore. High-speed connectivity eliminates the need for all but a small group to have permanent access to the data center. It's not unusual to have IT personnel working in a completely different building (or even geographic location) from their servers, and simply call a local IT shift technician to perform a hard reboot or load a tape as needed. Since this approach eliminates most traffic in and out of the data center, security effectiveness is enhanced by having a small list of people that the security guards will know by name and face. Any new face will instantly be detected and scrutinized- which is exactly as it should be.

While I have described two specific layers of access, there is a third level of access- to specialty areas, where support systems such as low-voltage electrical and mechanical systems are located. Only necessary critical-facilities engineering personnel should be able to enter those areas without an escort.

Depending on the safety systems installed, even more strict controls can be used to access high-voltage areas, where special personal protective equipment (PPE) is required.

SINGLE-PURPOSE FACILITY: Some managers and C-Suite executives like the idea of having their data center be something they can show off to others in industry, to demonstrate things like "Green" technology, IT innovations and the like. But there is a distinct problem with this, in that the security needs of a mission-critical facility are in direct conflict with such a technology "show-case." The data center should be as anonymous and "invisible" as possible, to avoid undue attention by those who might want to sabotage or otherwise harm the company. If miscreants don't know where your data center is, they can't harm it. Being anonymous and rarely visited by non-essential personnel, also means that it will be very easy to spot those who aren't expected.

From a practical perspective, it's important to remember that the more people are present within a facility, the more likely that unanticipated events can occur. For example, in my last data center, it seemed like there was someone burning a bag of popcorn in the microwave, at least once a week. In a critical facility cafeteria, that smoke and contamination can filter into the IT areas. People also do other silly things- clog toilets resulting in flooded areas, accidentally hitting the building with their car, causing fire alarms- or once, an actual fire- in the loading dock by trying to sneak a cigarette, etc. ALL of these events have occurred within data centers, and all are relatively innocuous- but all can have a negative impact on your critical facility. By minimizing the amount of personnel with access to the building, you reduce the risks of such events, with little cost.

From an engineering perspective, it's very important to remember that, no matter how much power you design your data center to handle, every watt is crucial to your operation. If you start with a building that is fed with 1 megawatt (mW) of capacity, and you use 200 kilowatts (kW) to support a lot of offices (air-conditioning, desktop computers, printers, desk-lights, overhead lights, AM-FM radios, cell-phone chargers, etc.), you've used up 20% of the *maximum* power your data center will have for the next 15-20 years, without earning a single dollar of revenue! Power and cooling in a critical facility are *extremely* expensive commodities. To expend those commodities on personnel, when you can accommodate them in a cheaper (and more comfortable) office building, does not make financial sense.

While the data center itself will be a single-purpose facility, there are a few areas that should be included in the overall building (though segregated from the data center itself, with separate security access protocols):

IT storage and staging areas: certain personnel within the data center, such as those who "rack and stack" new servers into their computer racks, remove old equipment and need some amount of storage, for day-to-day operations. Such areas typically have at least some (sometimes a lot) of high-dollar pieces of IT equipment, so the potential for theft or damage due to unauthorized access must be minimized.

Disaster-recovery (DR) locations: these are areas set aside with computers and monitors, power and communications ports, for IT personnel to work from, if another critical site is unavailable or unusable. This can range from a small conference room with provisions for a few people, up to an entire wing of a building, with room for 100 or more. Such areas typically have IT equipment set up and ready to go at a moments' notice, and access needs to be restricted so that the equipment remains undisturbed, ready for use when needed.

Parts and tool storage areas: a data center must have sufficient spares and tools on hand for the critical facilities staff to handle most immediate emergencies, with parts that normally require a long lead-time being on the shelf, ready for use in an emergency. [While not every part with a long lead-time needs to be kept on-hand, those with either a severe impact if a failure occurs or where the probability of failure is higher, must be.]

Engineering shop: the critical facilities engineering team will occasionally need staging areas for assembling things like long power cables, piping or other activities which require an open area in which to work. As well, the engineering team will provide certain services such as plumbing repairs, floor-tile modifications and other jobs which will require the use of specialized shop equipment, such as welders, band-saws, belt sanders and other similar gear. Due to the noise and dust given off by the operation of such gear, a shop is the logical location to store and operate such gear, especially since it will be away from the highly sensitive smoke-detection systems found in a data center proper.

Meeting rooms: While at least one meeting room is necessary, minimize the amount of them in the facility. Remember, floor space in a data center is costly.

As a final note, the physical construction of the facility shouldn't just meet building codes, but *well* exceed them; in a natural-disaster event, you need to know your facility is still intact, safe to work in, and able to support your operational efforts.

SUMMARY:

By downplaying the appearance of the data center, isolating it from other buildings, and minimizing the amount of personnel who can enter the facility, physical security is maximized, and risks due to sabotage or terrorism are minimized. Combined with a robust surveillance system, the data center can be an easily managed, highly secure facility, while still being cost-efficient.

Minimize unused spaces, and build the facility beyond minimum standards.

7 ELECTRICAL SYSTEMS

In this chapter we'll obviously go over electrical systems. That includes main distribution, PLC control, and "break before make" architecture.

Electrical schematics are like a road map. A map will tell you where the cities and roads are, what you're likely to see, and potential problems that might come up, like driving through an area you would prefer to avoid. A road map uses a standardized set of symbols, to make reading easy and consistent. The key, of course, is knowing what symbols represent what things. So we'll take a few minutes to discuss the more basic symbols, and have a more detailed discussion of electric plant design after.

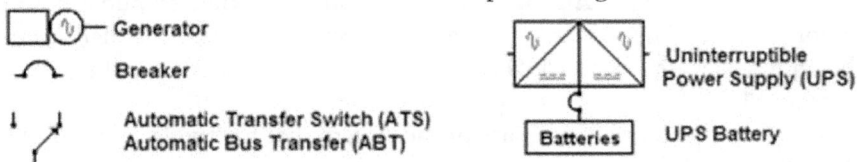

A generator is exactly what it sounds like (and we'll discuss them in depth later). The generator simply converts chemical energy into electrical energy.

A breaker is a large version of light-switch like you would see in your home. Some are manual, others are automatically opened or closed, and the size can vary from being small enough to fit in your hand, up to the size of a small car. Regardless of the size, the function is basically the same: allow power to pass through, or open when a fault is detected downstream.

An automatic transfer switch (ATS) is also commonly referred to as an automatic bus transfer, or ABT. As shown in the drawing above, the load side is on the bottom, and the arrow indicates which of two power sources are engaged. In this picture, suppose the feed on the right is considered the "primary," with the left as the secondary. If primary power goes away, the ABT will automatically switch to the secondary. When primary returns, the ABT will switch back to it, after a certain programmed time has elapsed.

For the UPS system, there are a variety of symbols used, with no apparent standardization at this time. Each UPS system will be adequately marked so that you can easily understand the map.

Now that we've addressed the map symbols, we're ready to discuss the Tier ratings of a site, as you may now be wondering what *exactly* the differences between the Tier levels are. Let's start with Tier-I, as shown in diagram 1, below:

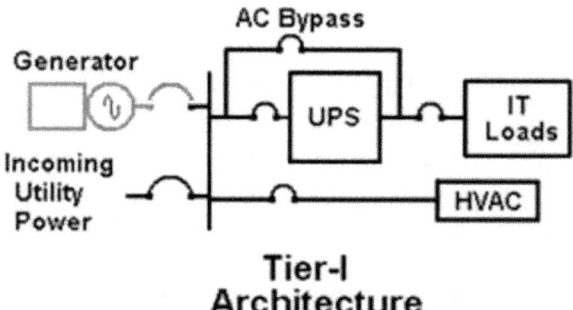

Tier-I
Architecture

In the above diagram, utility power is the normal source to the facility. Alternating current (AC) powers the heating, ventilation and air conditioning (HVAC). AC power is also fed into an Uninterruptible Power Supply (UPS). The AC is converted to direct current (DC), which keeps the batteries charged. The DC is then converted back to AC, and supplied to critical IT loads.

[In all of these drawings, parts of the system that have voltage are black. "Standby" or offline systems are in grey.]

You'll notice that you have a single generator to provide power if utility power fails. Of course, it may take the generator 10 seconds or more, to start. During that time, the Uninterruptible Power Supply (UPS) no longer has input power (which has already been converted to DC). Therefore, as the input voltage fails, and drops below the voltage of the batteries, the batteries maintain the voltage, like a safety net. This has the net effect of pulling DC power *from* batteries, where it is converted back to AC power, and supplied critical IT loads, until the generator starts. In this "double conversion" (AC to DC, and then back to AC) architecture, there is no delay in the battery supplying power- the effect occurs with zero time-lag.

There are several weaknesses with this system, however. To begin with, any maintenance that is to be performed on the generator or the UPS, exposes the IT loads to unplanned power outages, should the utility power fail while either the UPS or diesel is not available. Another weakness of this design is the presence of single points of failure, which will jeopardize continuous operation. In this example, a failure of the IT load breaker, the

HVAC feeder breaker or a short on the AC buses, which result in a loss of IT support (either power or cooling), resulting in unplanned downtime of the facility.

Clearly, while this design is relatively simple in design, it has significant weaknesses.

A key point to remember is that this is considered an "N" system. That is, the UPS has sufficient load capacity to supply all IT loads, and the generator has sufficient load capacity to support the UPS system and HVAC system.

In a Tier-II facility, some of the key weaknesses have been addressed, as shown below.

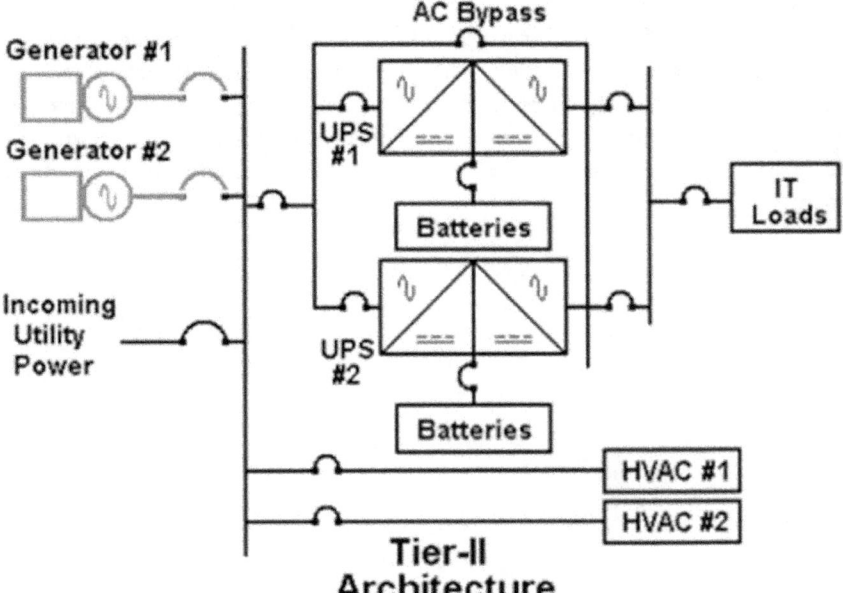

Tier-II Architecture

The presence of an additional generator, UPS module and HVAC capacity now reduces risks to the facility; if there is a failure of one of the generators to start, a failure of a UPS system or an HVAC system will not cause unplanned downtime of the facility. The doubling of the generators, HVAC systems and UPS modules, increases the costs, but there are obvious results; the redundancies of this system offer a much more reliable system than the Tier-I plant. This is considered to be an "N+1" system, meaning that if any piece of infrastructure equipment (such a generator) fails, there is an extra "+1" which can carry the load.

If IT loads equal 500 kilowatts (kW), and each UPS module is rated for 500 kW of power, then we have N+1 capacity of the UPS system. If, however, IT loads exceed 500 kW of power, then we are at N capacity, with no extra capacity to support the IT load, should a UPS module fail.

The same can be said for the generators; if total building load is 2 megawatts (mW), and each generator is rated for 2mW, we have N+1 capability.

That is, should a loss of incoming utility power occur, *and* a generator fails to start when this occurs, the other generators will have sufficient capacity to carry the building, without overloading (and possibly failing). The same is true with the HVAC.

Clearly, this system is starting to look very impressive. Yet, there are risks in this design due to single points of failure; a single IT power path and single AC buss (which both generators and the utility power feed), make this a Tier-II facility. It is good, but it could be *much* better.

A Tier-III facility, by Uptime Institute standards, is a facility where any piece of the building infrastructure can be taken out of service for concurrent maintenance. That is, an entire power (or cooling) path that supports IT equipment, can be removed for service, without interrupting operations.

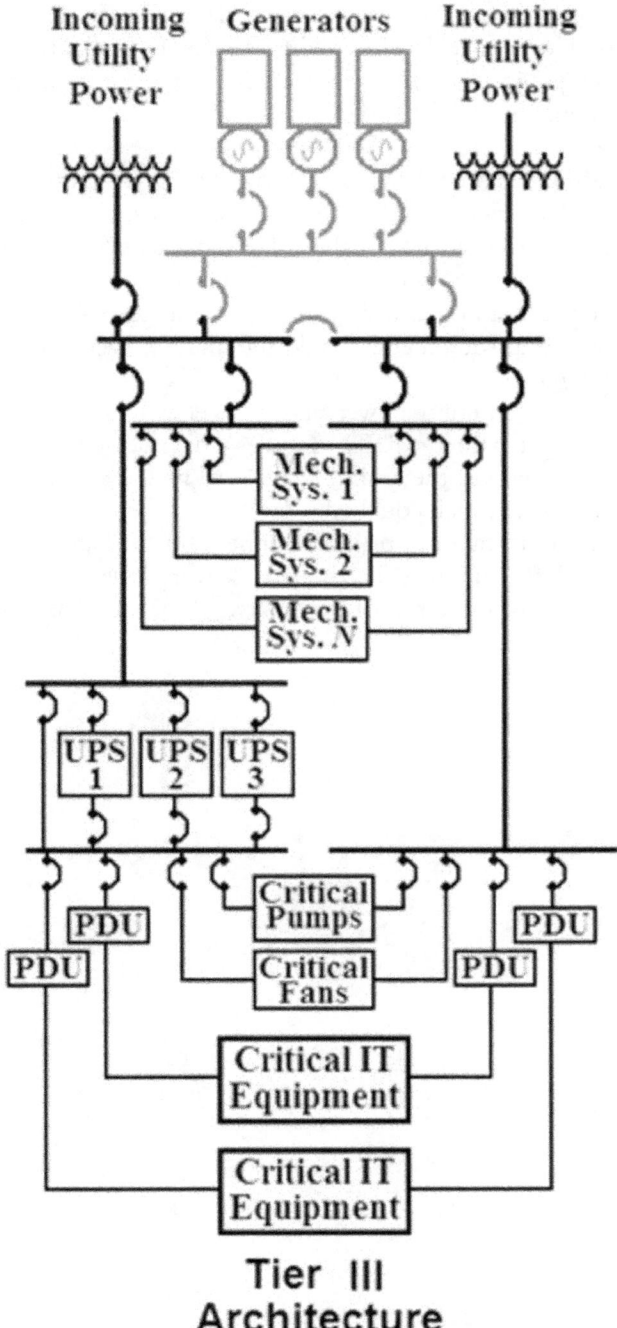

Tier III
Architecture

The drawing above shows a Tier-III electrical design. The system generally provides a solid level of redundancy and flexibility, allowing the service of all sections of this electric plant, while not interrupting IT equipment operation.

However, there is a weakness… Notice that UPS modules are only on the left side. The right side is effectively a "bypass" for critical power systems, allowing maintenance of those UPS modules (one or all of them), while utility power is directly fed to the critical loads.

The risk of such a design is that the UPS modules will not be available to power the critical equipment, should the utility power fail or go outside acceptable parameters.

If, for example, all utility power fails, the generators will start within 30 seconds or so, but with the UPS modules unavailable to provide power during those 30 seconds, the critical facility will go dark, necessitating a full IT systems recovery (which could take days).

Suppose your critical equipment consumes 300 kilowatts (kW) total, and each of your UPS modules is able to deliver 100kW. Consider the implications of this in terms of reliability. We'll discuss this question shortly.

A Tier-IV data center, is the ultimate in building infrastructure. Put simply, all infrastructure systems are fault tolerant. That is, the system can sustain a failure in *any* component, without jeopardizing the continuous operation of the facility. While this statement seems very basic, in reality it is very difficult to achieve, and requires a level of sophistication that is rarely built into other buildings.

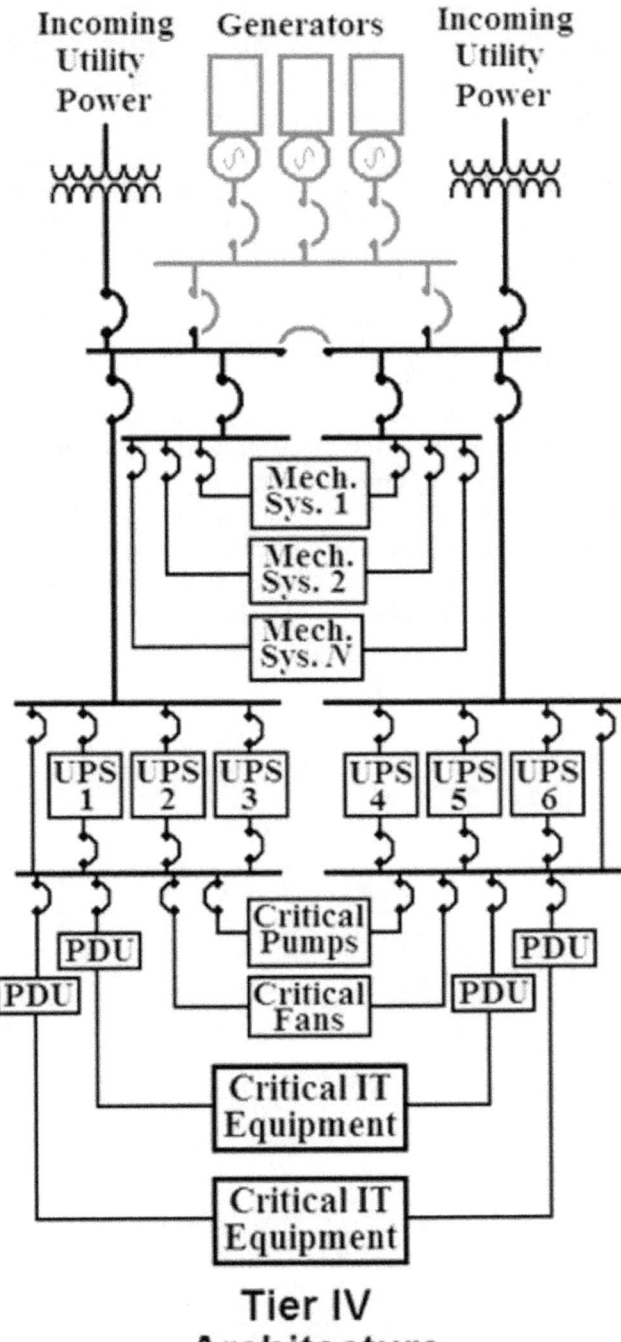

Tier IV Architecture

An example of a Tier-IV design is at above.

It would seem, upon superficial examination, that there is little difference between Tiers III and IV. But when you compare this design to the Tier-III facility, you will see that there are UPS modules on both sides of the plant.

While the Tier-III facility allows you to perform maintenance on any part of your data center infrastructure, without service interruption; a Tier-IV data center *assumes* that parts of your system will suffer unplanned events at the worst possible time, *and* in unexpected ways. If you have one side of your critical power system down for maintenance, the critical systems will be fully supported even during a utility outage, and the risks of interrupted service is effectively eliminated.

A key element that is almost never talked about, to make the Tier-IV system robust, is the UPS architecture: the "2N" system is considered the standard, but even this presents some problems.

When we talk about "N" systems, the N merely designates the number of units needed to support a particular load. If an airplane requires at least two engines to fly, but you have three engines, than the configuration of that plane would be "N+1," N equaling the two engines, in this case.

This is an "N" system. That is, if total IT loads are projected to be 100 kW, then the UPS is rated to 100kW as well. The UPS AC bypass is to allow maintenance of the UPS system, while still allowing power to be fed to the Power Distribution Units The diagram shows normal power going from utility power through the Auto Transfer Switch, the UPS and down to the loads. The generator and UPS AC Bypass breakers are shown in the open positions, to avoid confusion.

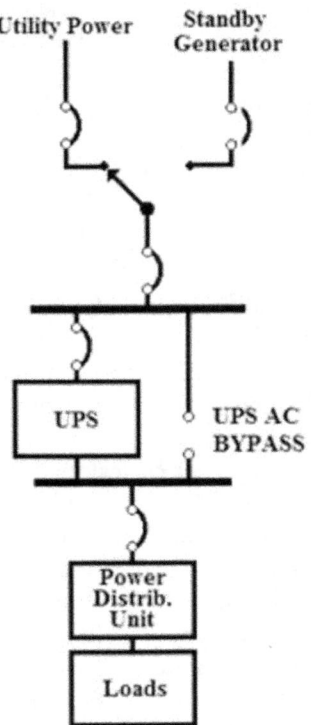

Advantages:
1. Optimum efficiency, since UPS load = 100%.
2. Theoretically simple, easy to understand.
3. Scalable- if loads increase, most UPS systems can be expanded for more capacity

Disadvantages:
1. Limited protection if UPS breaks down, or is offline for maintenance (and utility power fails).
2. Many single points of failure.

One aspect about this particular design, is the Automatic Bus Transfer (ABT) that switches between utility power and the generator. The put it precisely, the ABT is actually two breakers that are mechanically tied together, so that when one is open, the other is closed.

As long as the mechanical linkage between the two remains intact, the ABT is a very robust piece of equipment. The danger is that the mechanical linkage sometimes fails, and allows two independent power sources to connect to each other in an uncontrolled manner. This results in catastrophic failure, usually involving a fire, and sometimes arc-flash explosions. Doing a quick google search the terms automatic transfer switch failure explosion, I came up with more than 1.5 million hits.

The takeaway from ABTs is this: they work well, until they don't work.

When they fail, it's usually VERY bad.

This is an "N+1" configuration of UPS modules. Notice the load is 100kW, and each of the three UPS modules is rated for 50kW of power, and they are sharing the load. In an ideal situation, you could actually load 150kW total, on the UPS modules. However, you want to avoid directly feeding your critical loads directly from utility power, where harmonics, frequency and voltage variations could disrupt IT applications. In this system, failure of a single UPS will not force the critical loads to use utility power.

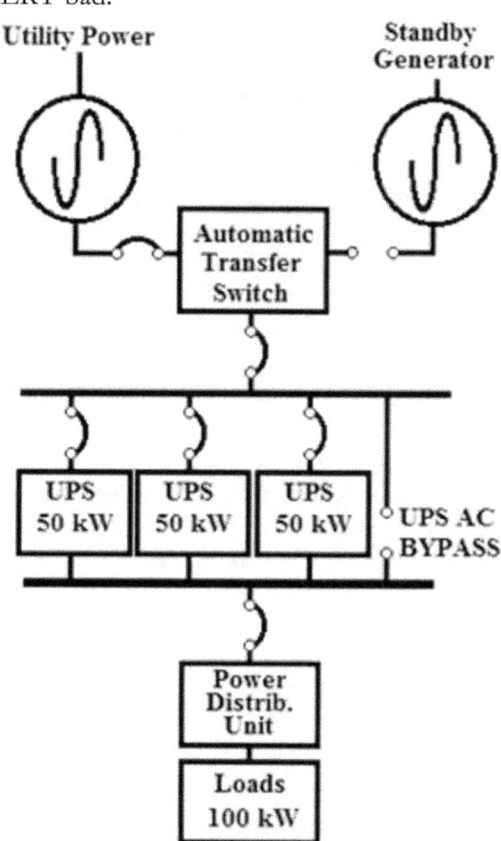

In normal situations, you can take a single UPS module out of service for maintenance, while still fully protecting your loads.

You still have single points of failure within the system- the breaker between the AC input buss to the UPS system and the Automatic Transfer Switch (ATS), the ATS itself, and the breaker feeding the power distribution unit.

While your single points of failure still exist, the most likely failure- a failure of a UPS module- has been effectively eliminated.
Advantages:
1. Mitigates danger of single UPS module failure.
2. Allows concurrent maintenance of UPS system without relying on utility power.
3. Allows additional UPS modules to be added as needed, if critical

load increases beyond forecasts (almost always happens!)

Disadvantages:

1. Single points of failure still exist.
2. If a loaded UPS module fails while an offloaded module is being maintained, utility power is the only safety net.
3. Maximum UPS efficiency (ignoring internal losses) =66%. If four modules share the load, then max efficiency equals 75%.

This is a "2N" UPS system, along with a Tier-III electrical architecture. Multiple utility power paths are available, but only one is active. In a configuration like this, a Programmable Logic Controller (PLC) would operate the breakers.

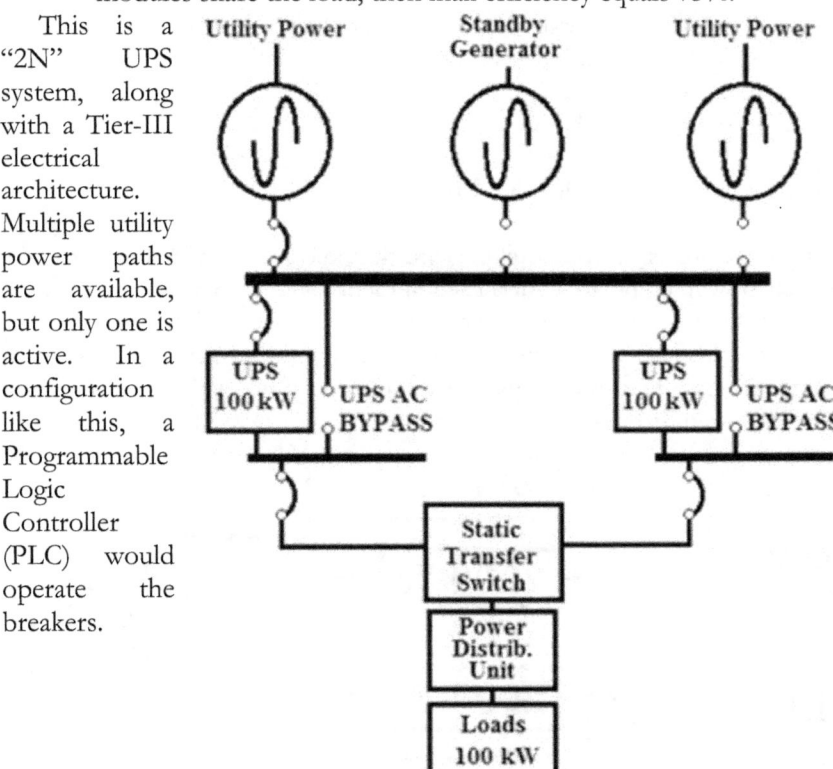

If the prime utility power source was lost, the PLC would open that breaker, and close the reserve utility power breaker. If that failed, then both utility power breakers would open, the generator would start, and the generator breaker would shut.

For the UPS system, the output of both UPS "loops" would feed into a static transfer switch that could select between either loop. If the UPS that was carrying the loads failed, the static transfer switch would shift loads to the other system within 4 milliseconds- short enough that critical loads will never see the transition.

Advantages:

1. Multiple power paths available
2. Allows concurrent maintenance of UPS systems, utility feeds or generator, with minimal risk to data center.
3. Allows additional UPS modules as needed, to match increasing loads.

Disadvantages:
1. Static transfer switch becomes most likely failure-point.
2. Relatively low efficiency of UPS systems (50%) since one loop is in standby at all times.
3. Initial cost very high.

Let's go into the UPS loading a bit more...

If the IT loads in your building are 400 kilowatts (kW), and you have dual-fed power going to your IT equipment of 2 UPS modules on each power loop, with each module capable of delivering 200kW, than the system would be "2N;" that is, it offer two separate feeds to your equipment, each feed fully capable of supporting the load. An illustration of such a system is below:

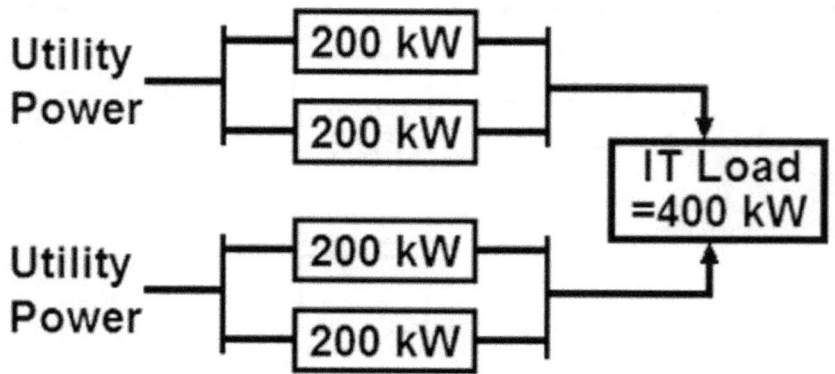

If your UPS modules are rated at 250 or even 300kW, it would still be "2N," simply because one side could support the load. However, suppose you are performing maintenance on the upper UPS loop so it is unavailable. The diagram would therefore be the following:

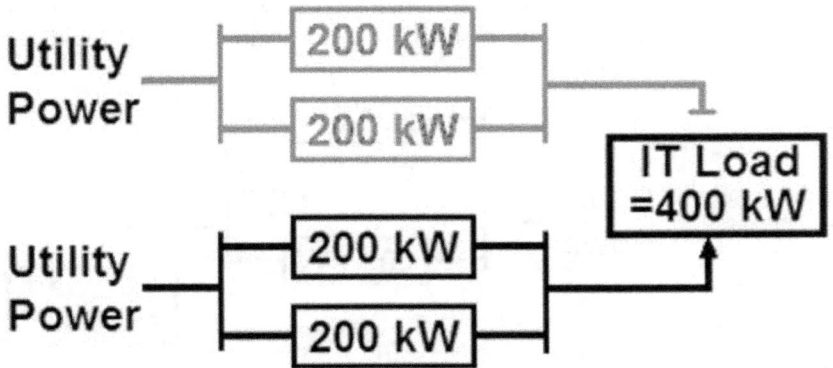

All of your eggs are now in one basket- in this case, the two 200 kW modules are still carrying the load. But either module from the lower loop fails, the remaining module will overload and trip offline (akin to the breaker in your home), and the data center will fail completely. Again, even if each module was rated at even 300 kW, as long as the remaining modules in a single loop are less than the total IT load, it's game over. And don't kid yourself- if one loop is unavailable for an extended period of time, this risk becomes *very* real. It's happened on my watch where my customers were hanging off of one loop for six weeks, and the UPS modules on the remaining loop were already beyond end-of-useful-life criteria. Believe me when I say that such a situation made for a lot of sleepless nights. As well, when failures occur, they usually will happen at a time when your safety net is weakest, and will cause the most damage. I don't know *why* that's true, but it is, nonetheless.

So, all that being said, how can you protect yourself from such an eventuality? I always recommend- for the ultimate in reliability and

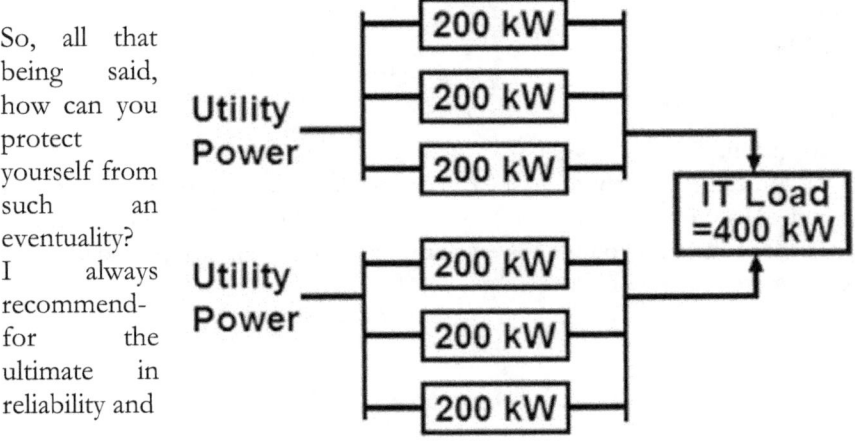

redundancy- that your UPS configuration always be "2N+1." Using the 400 kW IT load and using multiple 200 kW UPS modules, we would have the following configuration

In this case, if the IT loads are hanging off a single loop for maintenance, it would look like this.

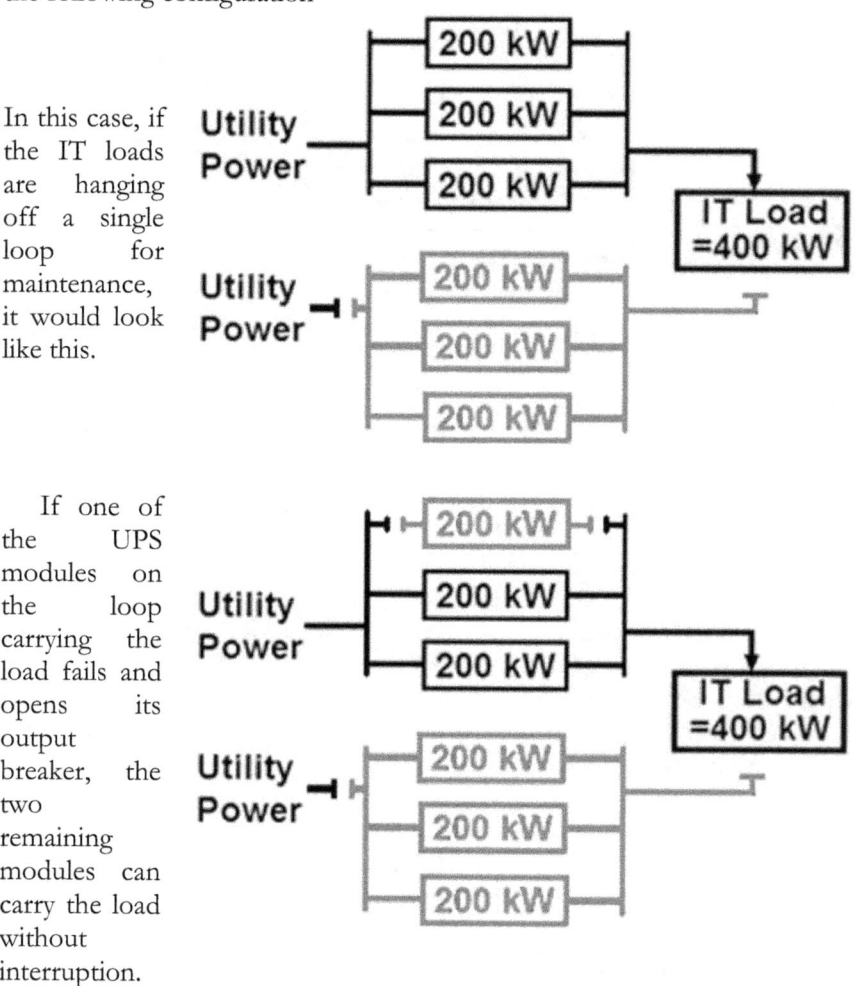

If one of the UPS modules on the loop carrying the load fails and opens its output breaker, the two remaining modules can carry the load without interruption.

If you cannot accept at least 12 hours of downtime every six months (or a single 24-hour period per year), then you should consider a 2N system. If you cannot have any outages even during such maintenance periods, then a 2N+1 system is the only way to go.

8 ELECTRICAL THEORY AND UNINTERRUPTIBLE POWER SUPPLY SYSTEMS (UPS)

ELECTRICAL THEORY AND UPS SYSTEMS

To begin with, we need to start with an introduction to basic electrical theory, and how it applies to motors, generators and uninterruptible power systems (UPS). The reality of *any* critical facility, is that 75% of what you need to know, revolves around electrical power. If you have this piece of the equation mastered, the rest will be relatively easy.

To understand how electrical systems work, we have to define some common terms, which may seem obvious at first glance, but have significant meaning in the explanations which follow.

Voltage is the electrical "potential" to deliver electrical energy. Its symbol is usually an "E" or a "V." If you were to think of it in everyday terms, it would be similar to pressure in a water pipe. Current (commonly symbol is "I") is a measure of electron "flow," or how fast the water flows through the pipe. Because current measures how much electron flow is actually occurring, you must remember that this flow generates friction, which in turn generates heat. It is important to remember that the friction goes up as squared function of the amount of current. This is commonly referred to as "I^2R losses," and this is why an overloaded wire can burn up!

Resistance, as the name implies, is the difficulty with which electrons flow through a system. Since we've been using a water-pipe as a comparison, this would be how rough the inside of the pipe is. The rougher the inner walls of the pipe are, the more difficult it is for water to flow through, and the lower the pressure will be at the end of the pipe. Make sense?

The chart at the right, is commonly referred to as "Ohm's Law," which lists the rules for understanding the interrelationships between voltage, current, resistance (in Ohms) and watts (actual amount of power delivered). Suppose your UPS battery has 480 volts of power, and you're charging the battery system with 6

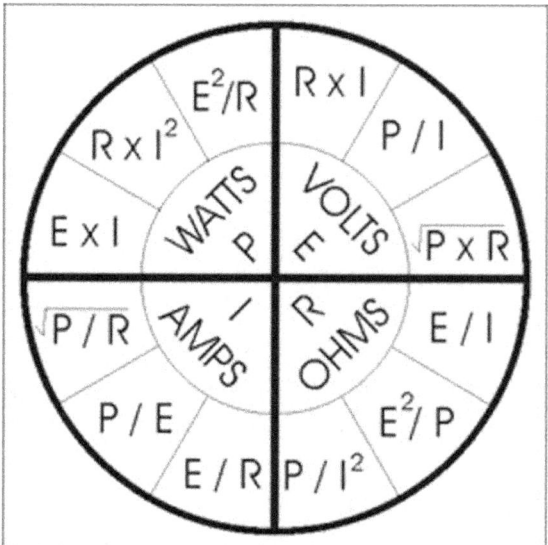

amps. That means the total amount of power being sent to the batteries, ExI, is (480 x 6 amps = 2880 watts), ~2.9 kilowatts, of power. We can also use this equation to determine the total electrical resistance across the battery, by using E/I, or 480/6 = 80 Ohms. Using Ohms' Law is one of the first tasks any electrician learns, and he will use it throughout his career.

There are two types of voltage, which you see all around you, every day. To begin with, we have Direct Current, or DC systems. DC offers the ability to store energy, in chemical form, like in the batteries for your child's toy, or the battery in your car. The significance of this, is that as batteries store more and more energy, the voltage between the positive and negative terminals increases, and it will begin to discharge (supply power) to the electrical system it is connected to, whenever the system voltage falls below that of the battery.

For example, in a car, if you happen to stop at a traffic light, with your air conditioner blowing at high speed, with your headlights on, the overall load may exceed the charging system of the car; it is consuming more energy than the alternator (which is a small generator) can provide. As the alternator is overloaded, system voltages starts to decrease in magnitude, until it falls below that of the battery. At that time, the battery stops the voltage from falling further, and provides power to the electrical system. In effect, that battery *stabilizes* the electrical system, by supply additional power whenever it is needed. Were you to look at DC voltage, it would simply be a flat line, at whatever voltage the battery happened to be at. Unfortunately, it's very difficult to transmit DC voltage over long distances, due to frictional (current) losses. As well, it's very difficult to convert DC voltage to high levels, to get around this problem.

The second type of voltage, commonly referred to as "Alternating current," or AC for short, is the type of electrical energy commonly found in your home, and most electrical systems around the world. It has an advantage in that it can be readily converted to very high voltage levels, and transmitted across long distances.

If you were to look at AC voltage, it actually swings from positive to negative and back, in a sine-wave form (this will be important later). While we won't go into the details of how it is accomplished, the AC waveform allows the use of transformers, which can "step" the voltage either up or down, with minimal losses of energy. With power (as shown in the chart above) being equal to voltage times current, if we double the amount of voltage, we halve the current flow, which in turn reduces heat to ¼ of what it was. This is why power lines that supply towns and cities are always of very high voltage- to maximize efficiency, and eliminate unnecessary heating of the cables.

A motor, in its most basic form, is a machine that converts electrical power (voltage times current) into mechanical power of a rotating shaft. Since horsepower is defined as (torque times revolutions per minute)/5252, you can see the direct correlation of electrical power (watts) to mechanical power (horsepower). A generator does the exact opposite of a motor- it converts mechanical energy into electrical energy. It's important to remember, than a machine can be a motor or a generator at different times, without any need to change the physical construction of the machine...

Now, you're probably asking yourself, what does this all mean? The simple answer is that the different types UPS systems we'll discuss, all try to do the same thing: provide stored energy to critical loads, until backup generators can start up, and take over. When you keep this context in mind, you'll find that it's actually pretty simple to understand.

The Motor-Generator Set

The most rudimentary type of Uninterruptible Power Supply (UPS) system around, is the "motor-generator set," or MG. They have been around, in various forms, for about 100 years, and were initially designed to provide power to the earliest submarines, before World War I.

In simple terms, the MG set connects the AC (normal) electrical system to the DC (stored energy) system *through a rotating mechanical shaft.*

These systems are essentially alternating-current AC motors tied to direct-current motors, by a common shaft. This arrangement is commonly referred to as a "motor-generator set," or MG. They are still commonly used on submarines, and in some critical facilities applications.

While the following explanation may sound a little confusing, there's a simple way to understand how an MG set works: where voltage is higher than inside that side of the machine, that side is the motor; where voltage is below that of the machine, that side is the generator. Use the diagram on the next page, to follow along.

How an MG works is simple: each side of the machine can act as a motor or a generator. Let's start with the AC side of the machine: if the voltage of the electrical system that the MG set is connected (buss voltage) is higher than what exists on the MG, then that side of the MG acts as a motor, converting the higher level of electrical energy into mechanical energy, and turning the shaft, just like any electric motor would. On the DC side of the machine, the shaft is turning, where buss voltage is below the voltage inside the DC side, and this side of the machine becomes a generator. In a typical situation, the DC side would be supplying the extra voltage to batteries, which in turn convert the electrical energy into chemical energy, which can be stored for use at a later time.

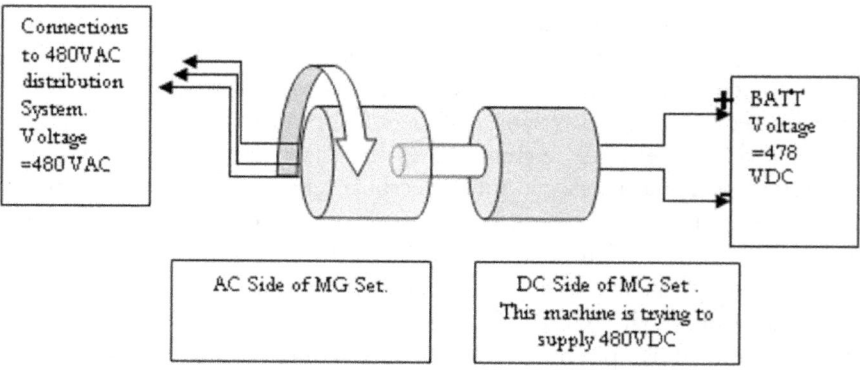

Suppose we're in a typical situation, as you'd see on a submarine... The boat is travelling through the water, power is being supplied by large turbine generators (TGs). On the other side of the system, there is an MG set. The buss voltage supplied by the 480 volts of alternating current (480 VAC). The MG set is acting as an AC motor/DC generator. That is, the AC voltage is driving the AC side as the motor, and the common shaft spins the DC side of the machine, as a generator, where the buss voltage is 480 volts of direct current (480 VDC).

Now, suppose there is an emergency, and the turbine generators are suddenly unable to supply power. In this situation, the system looks like the following:

As soon as the power supplied by the turbine generators begins to drop to a lower magnitude, the AC side of the MG set begins to slow down. As soon as the DC side of the machine supplies voltage less than the voltage across the battery buss, the battery will maintain the buss voltage. As the voltage supplied by the DC side of the MG falls *just* below the voltage supplied by the battery (which will maintain a constant voltage level), the battery will begin supply power to the DC side of the MG set.

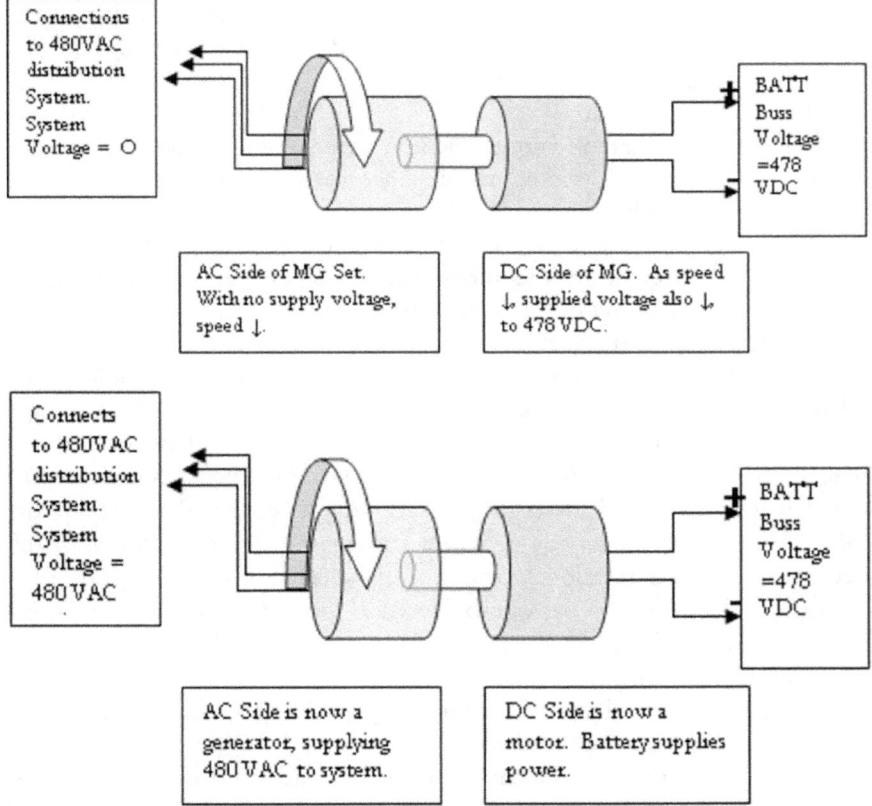

In effect, power flow has reversed, and now the battery supplies the MG, making the direct-current side of the MG set a motor. Now, the slowdown in speed of the MG shaft has stopped, because the DC side of the motor is powering the shaft. As a result, the AC side of the machine, because it is no longer receiving a supply of power from the turbine generators, supplies voltage that is higher than buss voltage; it is now a generator.

While this explanation seems quite complicated, the reality is that it isn't. If you keep in mind that one side always acts like a generator, and the other acts like a motor, and that their role is determined by buss voltages, then the rest becomes easy.

There are some advantages to MG sets.

1. They're proven technology, and can be very robust in construction and reliability.
2. They are very compact, and require no special environmental conditioning.
3. An MG set can provide power for as long as the battery lasts, which can be minutes or even hours, depending on the capacity of the battery.

The tradeoffs are that

1. They aren't very efficient in terms of energy utilization,
2. require periodic maintenance for things like motor brushes and bearings,
3. Have a relatively high total cost of ownership, and
4. The batteries required require a significant amount of space and environmental control.

Flywheel Energy Storage

The next type of emergency backup system uses 1950's technology in a relatively new approach, and is commonly referred to as "Flywheel Energy Storage" technology, or FES. An FES system rotates a flywheel to very high speed, and uses the rotational energy. As more energy is delivered to the flywheel, the speed increases. Conversely, as energy decreases below that of the flywheel, as the flywheel provides its stored energy, speed decreases. As the rotational from the flywheel is consumed, it is then converted electronically back into electrical power, to provide a short-duration power source to protected loads.

Advantages:

1. They can have an exceptionally long useful life, with very little maintenance required.
2. The only inefficiencies of such a system are mainly due to frictional losses of the flywheel spinning in the air (minimized by being in vacuum chamber) and bearings, thus offering exceptionally high levels of electrical efficiency (up to 98%).
3. They are not sensitive to environmental conditions, and can be installed in a relatively small footprint (as compared to other UPS systems that require batteries).

There several disadvantages of FES systems, and they are highly significant to their appropriateness in a critical facility.

1. The energy storage of the flywheel is limited by the tensile strength of the flywheel itself. Once the flywheel rotational

forces exceed the tensile strength of the flywheel, the flywheel will literally tear itself to pieces, in an explosive manner, releasing shrapnel into the vicinity.

2. The second issue, is that FES systems can only provide electrical backup for a very small amount of time- typically, 15 seconds at maximum.

3. A Tier-IV facility requires an electric plant which is described as "self-healing." In other words, if a cascade failure begins, the plant needs sufficient reserve power in the UPS system to allow enough time for the plant to detect the cascading fault situation, stop the cascade and put the electric plant in a safe condition. This requires minutes of reserve power in a UPS system, typically 15-30 minutes. Because a flywheel UPS system offers reserve power measured in *seconds* it cannot, by definition, be utilized in a Tier-IV electric plant.

4. A final negative aspect of FES systems is that they're riding on bearings, and the flywheels are in a vacuum (to reduce aerodynamic drag), which means the bearings wear out, and the vacuum pumps wear out as well. When I did a total cost of ownership (TCO) analysis of flywheels versus typical double-conversion UPS systems (including the accounting of real-world electrical loading) the FES systems were more expensive than UPS systems that are used throughout the data center industry.

Line-Interactive UPS System:

This type of UPS system, from a historical perspective, has been deployed more than any other, because of its durability and low cost. Unfortunately, it can used with loads up to 5000 watts, as a practical limitation. A line-interactive system actively filters incoming power from disturbances including voltage spikes, using passive filters and a tap-changing transformer. If incoming power falls below specified parameters, the UPS uses an inverter to supply power within specifications. When the inverter takes over, it disconnects from incoming power, and uses the batteries for energy.

In areas where utility power tends to be unstable- wild swings in voltages, high-frequency noise or harmonics, the line-interactive UPS has limited abilities to correct such variations, and may have to rely on battery power frequently, sometimes several times in a single day. And while this may or may not fully discharge the batteries, it will certainly decrease the useful life of the batteries, necessitating more frequent replacement, and therefore, increase total cost of ownership.

A very good aspect of this approach, is that the line-interactive UPS system is very energy-efficient, as it is passive in most situations, and only uses power to keep the batteries fully charged.

Advantages:

1. Energy-efficient UPS topology, up to 98%.
2. Fewer components than double-conversion UPS systems,
3. Less heat generated UPS areas,
4. Lower initial costs
5. Lower utility costs

Disadvantages:

1. Only useful with loads of <5000 watts
2. While fewer components, they are much bigger, thus typically larger, heavier
3. No ability to correct power factor
4. Not suitable for areas where utility power is highly unstable

Double-Conversion UPS:

Like its name states, the double conversion UPS system takes its incoming AC power, converts it to DC (where some is used to charge the batteries), and then reconverted back to AC. This approach is ideal for critical loads exceeding 5000 watts, which is most data centers. This topology tends to use many more components than the line-interactive UPS design, but the components are all much smaller, resulting in a smaller and lighter overall package. However, the high number of components results in higher heat generation, which must be cooled more aggressively than the line-interactive design.

Because a double-conversion topology always converts all power at all times, this results in electrical inefficiencies that the line-interactive design does not suffer from.

A double-conversion UPS converts AC energy to DC energy, which is then used to maintain full charge on the battery. The UPS then converts it back to AC, using a series of switches, to vary between minimum and maximum voltages. This is commonly referred to as Pulse Width Modulation, or PWM. For example, in the following picture, PWM varies between -1.3 and +1.3 volts. There is a delay, where the voltage of the system has to "catch up" to the maximum magnitude of voltage. Because of the delay, if the pulse is a very narrow of window in time, the buss voltage will only get a certain percentage of that magnitude. And the pulse time gets longer (wider pulse "width"), higher voltage occurs. So, by varying the amount of time between each switching event, the overall average voltage increase or decreases, as needed, and you can now create an AC waveform, which is similar to the AC input.

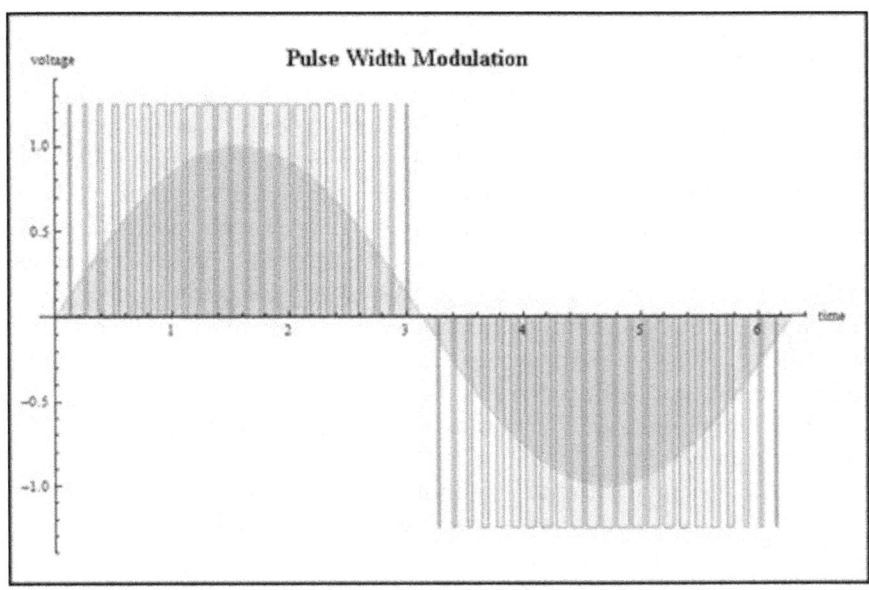

There are several advantages to this approach, which are very important to the continued operation of a critical facility.

First, by converting all incoming power from AC to DC and then back to AC, you eliminate nearly all incoming power distortions, such as harmonics, which can cause havoc for sensitive electronic equipment.

Secondly, if incoming power falls below what would be considered acceptable levels for a line-interactive or flywheel system, the double-conversion UPS can still utilize incoming power, and correct the voltages to whatever magnitude is desired. The same is true for frequency variations (which occurs in utility power supplies, if the grid becomes overloaded) - the line-interactive system can use incoming power that would be unacceptable for most electronics, and "clean" it up, so that the protected equipment sees none of these variables.

Advantages:
1. Reasonably energy-efficient UPS topology, from 92-97%, depending on design and loading.
2. Excellent for areas where utility power is highly unstable
3. Corrects Power distortions
4. Adjustable frequency and voltage specification
5. Filters out dangerous harmonics
6. For loads >5000 watts

Disadvantages:
1. More heat generated
2. More components than a line-interactive UPS system

3.   Higher utility costs

"Delta conversion" UPS System:

A recent addition to the UPS market is the "delta conversion" system. It is touted as highly efficient while automatically correcting voltage distortions.   After working closely with a delta conversion UPS and examining the schematics, it is essentially a scaled-up line-interactive UPS, using a "buck-boost" transformer to correct voltage almost instantaneously.

Unfortunately it suffers from the same weaknesses of its older brethren; inability to correct for frequency deviations and the allowance of high-frequency harmonics (noise) to pass through the system. I have personally seen them fail to properly switch from line-interactive mode to battery mode, due to a variety of control issues.

The overall results are the same as smaller line-interactive systems:

Advantages:

1.   Energy-efficient UPS topology, up to 98%.
2.   Fewer components than double-conversion UPS systems,
3.   Less heat generated UPS areas,
4.   Lower initial costs
5.   Lower utility costs

Disadvantages:

1.   No ability to correct power factor
2.   Not suitable for areas where utility power is highly unstable

Other considerations of UPS topology would include those designs that attempt to reduce power consumption of internal components, increasing operational efficiency.   Typically referred to as "eco" models, they use variations of attempting to remove some components to the system and re-engage them when input power is either lost or out of tolerance. Remember though, that re-engaging those offline components takes time… and moving parts.   And this means that reliable operation can be compromised due to either increased wear or unexpected control faults.

UPS Energy Efficiency: All Sizzle, No Steak

With the current trend to "go green," everyone is interested in better efficiency, right? So we can buy a more efficient UPS that'll still protect us, the boss will be happy that you reduced costs, and you'll still have superb reliability, right? Maybe, but maybe *not*.

UPS sales vendors love to tout the energy efficiency of their latest whiz-bang UPS system, because they know the "going green" movement is a great selling point for them. And just about every UPS manufacturer makes a "greener" model than the standard war-horse design. But is it really so efficient?

Let's look at a typical UPS architecture of a highly redundant UPS system in a facility I worked with:

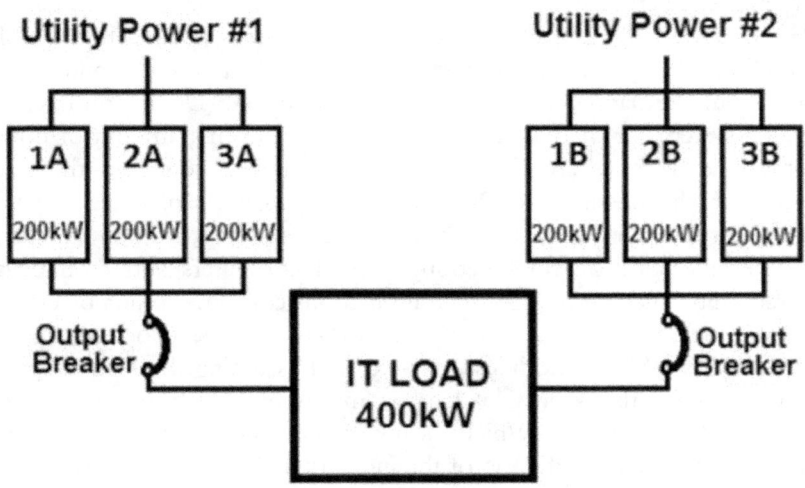

Each UPS loop has 600kW of power, or 1,200kW of combined power. However, we occasionally have to take a single UPS module (such as 2B) offline for maintenance, or even an all of the "B" modules, to perform maintenance on the breakers or conductors. To allow continuous operation of the IT equipment, therefore, UPS loop "A" will have to carry all of the electrical load. But you're thinking, "wait a minute, why are there 600kW of UPS power when I have only 400kW of total load?" The answer is that, occasionally, a facility will have to rely on a single loop for an extended period of time. If, during that extended time, one of the modules fails while running in single-loop mode, the remaining UPS modules can carry the load without interruption. This configuration is called "2N+1" because you have two independent UPS power loops which can carry the load, and each loop has one redundant module. (There is more on this in the section on electric plant design.)

We are constantly calculating the chances of some event happening. Gamblers and weather forecasters make their careers on such actions. But we rarely think such redundancy is needed in a facility. Yet in one data center I managed, a design defect was discovered in *both* of the output breakers, requiring the entire facility to be on only one loop for ~6 weeks while the breaker (and it's cabinets and support systems which were no longer supported by the manufacturer) were replaced. That particular facility was only "2N" with no redundant modules on each loop... and it was a very long and tense summer, while we were effecting repairs. Luckily, we made it through without any modules failing… but relying on "luck" is a really bad way to go.

Now, let's get back to the subject of electrical efficiency. In short, a UPS module consumes a certain amount of power to operate. In a double-conversion UPS, you have the rectifier which converts the incoming utility power from alternating current (AC) to direct current (DC). You have the charging system for the batteries, which consumes some amount of power whether the batteries are charging or not, and then you have the static inverter which converts the DC back to AC, as a clean, noise-free sine wave. You also have the controls for allowing the UPS to run in parallel to the other modules, instrumentation, internal computers and monitoring, data recording and so on. All of these functions consume some power.

Going back to the 2N+1 configuration we were just looking at, you have 1,200kW of UPS modules available, and assuming the load is spread relatively evenly, the 400kW of loading means each UPS module is only rated at 33% of capacity. Which means the peak efficiency of a particular UPS at full load may be the top of the industry, but what you really want to know is the efficiency at where your data center will be running. The graph below shows some efficiency curves from 45 different designs and models. As you can see, there are relatively inefficient systems out there, but most of them are all running in the same range of about 82-92% efficient, for the range we're looking at.

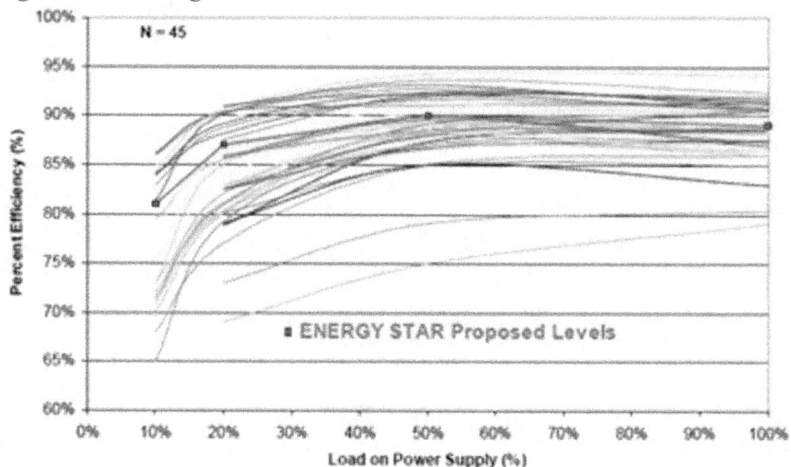

Graph from http://emidatacenterblog.files.wordpress.com/2010/02/server-ps-curve.jpg

You can be certain that the flywheel and delta-conversion systems will be on the highest end of the spectrum for a given load range, while modern double-conversion UPS modules will be in the middle of the field.

"That's all well and good," you may be thinking, "but what does this all mean?" We can convert the energy efficiency numbers into a cost analysis, so that you will have more meaningful information from which to judge what's right for your needs.

Suppose we have decided to replace all of our double-conversion UPS modules with delta-conversion units, which offer better electrical efficiency at the cost of reduced protection to the loads, both through time-delays when switching to battery-mode, and the inability to either filter out noise or maintain frequency. We'll continue on with the 2N+1 plant design.

At 400kW of IT load and 33% loading, we'll assume the new delta-conversion UPS modules are at the top of the efficiency curve with 92.5%. 400kW/0.925 = 432.4kW, so the UPS modules are consuming (432.4 − 400) =32.4kW

Let us also assume that the double-conversion UPS modules are in the middle of the pack, at 85%. Using the same calculations, we determine that the 70.5kW.

What's the cost difference? In the area where I live, the cost of electricity is ~$0.17/kW-hr. The double-conversion UPS modules consume 38.1kW more, and are (of course) running 24x7, so the cost per day is $6.48 cents per day more, $ 2364.11 per year, or $23641.05 over the full life of the UPS systems at ten years.

It would seem as if this is a "good deal," right? As always, "it depends." If the delta-conversion UPS system fails to switch to battery in time due to degraded utility frequency just once, and your IT equipment goes offline (in other words, you have an unplanned outage of your data center) the cost will exceed any savings you might have garnered, by several orders of magnitude.

If you're considering replacing a less efficient UPS system (that is still well within its life-cycle) with a more efficient, "green" UPS system, the same calculations can be used to determine if the energy savings will ever pay for the new UPS system. Indeed, I made these exact calculations, and the payback period due to energy savings was some absurd number like 300 years. (They decided not to do the switch.)

So, you can see that being "green" doesn't accomplish much, and can easily cost you far more in terms of both up-front money (as the expensive-to-maintain flywheels demonstrate) and more vulnerable to unplanned outages (as those same flywheels won't allow the facility to "self-heal" during cascading failures.

UPS Design Summary

I have personally seen all of the different designs described in operation, as well as having to provide a "safety net" when the rest of the facility was in serious trouble. If your goal is absolute reliability no matter what, then the only real choice is double-conversion UPS systems. Powerware, Liebert and Mitsubishi all have excellent offerings. Powerware- a relatively small player in the UPS industry recently purchased by Eaton, makes absolutely *superb* systems that will take an incredible amount of abuse and keep right on working. As well, their service providers are excellent, both in terms of responsiveness and technical ability. Given the choice of UPS manufacturers that are available in the market-place, these are the vendors I would most strongly recommend.

With regard to delta-conversion UPS systems, my experiences with them have been... less than stellar. With mission-critical facilities, half-measures and allowing certain electrical disturbances to pass straight through into your very expensive and very delicate equipment, is playing with fire. Sooner or later you're *will* get burned. I've never been burned by the mentioned double-conversion UPS systems, but I have by the delta-conversion systems.

As you have probably noticed, I have little respect for the flywheel technology; in my opinion, the amount of energy storage of a UPS system must be measured in tens of minutes. This relegates the flywheels to low-standard data centers, where politics and appearing "green" outweigh the practical realities of the engineering world.

Let me give you a real-life example of what can happen with flywheels...

At two data centers in Europe, owned by the same company on a large campus, their UPS systems were large flywheel systems. The local utility power came in through two separate feeds, one of which crossed over a river, along a bridge. As a large ship was navigating through the river, the bridge had to be raised, requiring the utility feeder cables to be disconnected before the bridge could come up. The entire region had their power rerouted to the alternate (non-bridge) feed, but this overloaded the alternate feed. As the overload continued, the utility power generating station started to lose control of frequency (the loads were literally dragging down the generators), and the frequency started to slow down. The flywheels in the data centers also slowed down, since they rely on constant frequency to keep them at sufficient speed. When the alternate utility feed finally failed on under-frequency fault, the flywheels in the data centers were not spinning fast enough, and had insufficient rotating energy to power the data center until the emergency generators started and came online. Both data centers suffered unplanned outages simultaneously. It was a result of improbably circumstances combining with insufficient reserve power, which lead to this operational nightmare.

Conclusions:

There is much discussion, and frankly, a great deal of misunderstanding about the function of UPS systems, how they work, what they offer, and even how to determine what is best for your application. A UPS system, configured properly, should be able to provide emergency backup power until replacement power comes online, usually through standby generators on your site.

If you're in an area where power is unstable, and you often have outages or voltage "sags," frequency problems or your neighbors operate robotic systems (which introduce dangerous harmonics), then avoid line-interactive or delta-conversion systems.

The specific goal of mission critical UPS systems is to provide a final safety-net to the clients' gear. They may be interested in "green" technologies, and other unique approaches to accomplishing a function. But don't be fooled; those clients are not going to be impressed when that shiny new technological marvel leaves their IT servers sitting in the dark for 30 seconds, and they have to spend the next 48 hours without sleep, trying to restore their systems and rebuild lost data. All the client is going to care about, is that the UPS systems carried the loads with absolutely *no problems.*

Build your facility, understanding this and choosing the most robust systems you can get your hands on. "Green" matters far more to the company, when that green is lost revenues because their servers were offline.

In my opinion- having experience with all the systems mentioned (except "flywheel" technology) - the best system to have, without question, is the double-conversion systems. When properly configured in a 2N+1 configuration (we'll get to that discussion later), this represents the premium level of safety for a critical facility.

Finally, don't just trust the salesman of the company that tells you their product is "the best." Sales people get paid for what they sell, so they will tell you whatever they must, to convince you to buy. If they told you their products weren't so good, would you buy them? They'd be out of a job!

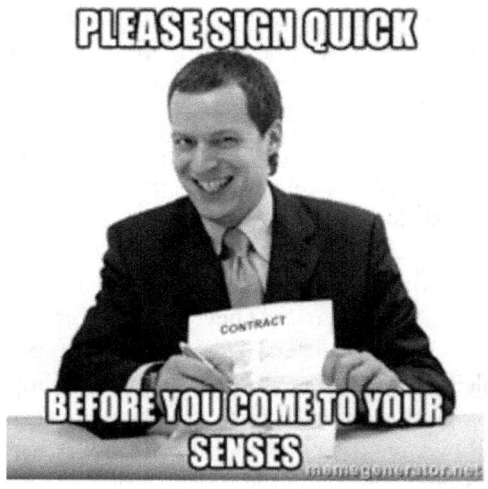

I had one UPS regional sales manager tell me "our units haven't lost so much as a <u>microsecond</u> of critical power for IT clients" the day after the new UPS he just sold us had failed, causing an IT equipment outage. He knew about the failure, and yet continued to say this. Upon further investigation, it turned out that this particular design had failed at a number of other facilities in exactly the same manner, and it was common knowledge within the manufacturers' organization.

I wish I was joking, but I'm not.

They'll "sell the sizzle," using all the latest buzz-words, fads and gimmicks, to get you to buy their equipment.

Armed with a better understanding of how UPS systems *really* work, you'll hopefully have enough information to make truly informed decisions that are in your best interests.

9 UPS BATTERIES

Since we have spent a fair bit of time on UPS systems and underlying operation, it seems only logical to include UPS batteries at this point. While their packaging varies quite a bit, the vast majority of batteries use lead and acid, to store electricity. The chemical equations are as follows:

Equation 1
Electrolyte
$$H_2SO_4 \rightleftharpoons H^+ + HSO_4^-$$

Equation 2
Negative Electrode
$$Pb_{(metal)} + HSO_4^- \overset{Discharge}{\underset{Charge}{\rightleftharpoons}} PbSO_4 + H^+ + 2e^-$$

Equation 3
Positive Electrode
$$PbO_2 + 3H^+ + HSO_4^- + 2e^- \overset{Discharge}{\underset{Charge}{\rightleftharpoons}} PbSO_4 + 2H_2O$$

Equation 4
Total Reaction
$$Pb_{(metal)} + PbO_2 + 2H_2SO_4 \overset{Discharge}{\underset{Charge}{\rightleftharpoons}} 2PbSO_4 + 2H_2O$$

The important point of the equations is with line #2, where 2 extra electrons are left available on the negative electrode during discharge (when the battery is supplying power), or the two electrons are being forced back through the positive electrode in equation number #3, when charging the batteries. Other elements can be similarly combined, but the end goal is the same; provide surplus electrons on one electrode for discharge, and reverse the electron flow to recharge.

Since lead-acid batteries (using the above equations) are by far the most common, so we'll discuss them with regard to packaging.

Flooded-cell batteries: To begin with, flooded cell lead-acid batteries oldest and most commonly used batteries (by far) throughout the world. As the name implies, the batteries has lead plates immersed in a jar that is filled with sulfuric acid and purified water. This approach is so common that almost any needed option can be found or added. Usually, there are multiple battery cells contained within a single "jar" to simplify manufacturing. The plates of each cell are typically mechanically supported by the top of the jar, where there is usually a fill hole and vent for hydrogen gas to escape.

There are some advantages of this type of battery:

1. Widely used.
2. Very dependable.
3. Most economical in terms of cost per kW of capacity, making this very cost-effective.

Some of the disadvantages of flooded lead-acid batteries include:

1. Delicate; the lead plates are supported by the top cover
2. Can crack or come unsealed, causing acid leaks
3. Extremely heavy
4. Not as durable as nickel or lithium-based batteries
5. Requires longer to fully charge after batteries discharged.

Sealed lead-acid batteries. In the attempt to avoid loss of water or gases from flooded lead-acid batteries, new "maintenance-free" sealed batteries were developed in the 1970's. Because they are sealed, hydrogen and water vapors are unable to escape, and recombine, avoiding the loss of water or acid within the cells. This in turn allowed the development of the valve-regulated lead acid (VRLA) and absorbent glass mat (AGM) battery designs.

VRLA batteries use a silica-based gel to hold the acid; therefore if the VRLA jar sustains mechanical damage, such as from a crack, the electrolyte will not leak out and the battery will continue to function. AGM batteries hold the electrolyte in a glass mat; the matting acts as a wick to prevent any leakage, and is also resistant to mechanical damage.

Advantages of VRLA/AGM batteries are

1. High power capacity versus physical size
2. Can operate in nearly any physical orientation
3. Maintenance-free
4. Insensitive to vibration
5. Can operate after sustaining significant physical damage

VRLA/AGM batteries do have some significant drawbacks:

1. Temperature sensitivity; as temperature increases, life degrades quickly
2. Very expensive per kW of power delivered.

Summary: Gel-based batteries offer enhanced safety, superior power deliver for a given volume of batteries within a building, and are more forgiving to the abuses that data centers sometimes dish out onto batteries in general, such as unexpected momentary interruptions of utility power. The biggest thing that gel-based batteries have going for them, is the safety aspect. They just don't leak, and therefore the fluidic hazards associated with flooded batteries, simply don't exist.

However, most fire-marshals generally do not recognize any inherent differences between flooded batteries and their gel-based counterparts, and therefore demand the same safety protocols for gel-based batteries that are used on flooded-cell batteries. This in turn brings about additional costs that are completely unnecessary.

If fire codes were able to recognize the lack of need for additional ventilation and spill containment, there would be significant cost-savings (by not installing ventilation and spill-containment measures) on top of the operational advantages to using the gel-based batteries, I would recommend using them. However, unless the local authority (fire marshal) is willing to waive the spill-containment and ventilation protocols that are required for flooded-cell batteries, the much cheaper costs of flooded-cell batteries seem a much better approach.

Life span: typically, battery manufacturers will specify a lifetime of 7-10 years, and some go so far as to claim their batteries can provide useful power for 20 years! As usual, there's a catch: they'll live that long, in an ideal environment. And "ideal" includes perfect ambient conditions, periodic discharge and recharge cycles (with an "equalizer" charge thrown in as needed), and so on. In other words, batteries are like muscles in the body; if you don't use them regularly, they tend to atrophy. In a mission-critical environment, you rarely have the opportunity to perform a full discharge/recharge operation, because of the very nature of the facility and its tenants. Therefore, the decades-long life that some manufacturers claim, simply will not materialize in the real world. In a mission-critical facility, you must plan on replacement of the batteries around years 5-7, as a policy. Better to have the money allocated and not use it, than to need the money, and have none available.

Battery Capacity

Battery capacity- that is, how long the batteries can sustain a given load- is directly related to two functions; cost and/or size.

You can obtain battery strings of high capacity levels in a relatively small package (such as gel-cells), but they are expensive. On the other hand, you can obtain equal levels of capacity using flooded cell lead-acid batteries for a lower price, but they will take up a significant amount of space in costly engineering spaces. Thus the trend is to minimize the amount of capacity available, to minimize both space and cost considerations.

But this logically leads to the question, "what is the optimal capacity?" To this, there is *considerable* debate in the industry, as to how much is "enough" capacity.

There are several distinct viewpoints:

1. Have sufficient capacity until the generators start (30-45 seconds). Make sure there is always at least N+1 capacity of the generators, and if the generators fail to assume the load, insure that IT operations will roll over to a colocation facility.

     a. Advantages:
          i.   Minimizes the costs of UPS batteries
          ii.   Minimizes space used by UPS batteries
     b. Disadvantages:
          i.   Precludes Tier-IV performance of electric plant
          ii.   Requires seamless rollover of IT operations
          iii. Higher cost of IT infrastructure
          iv.   Requires multiple data centers for effective coverage
          v.   Requires full data mirroring of data centers (i.e., <100 km distance between rollover sites)

2. Have capacity of between 5-15 minutes (medium capacity levels)
     a. Advantages:
          i.   Allows for Tier-IV performance of electric plant
          ii.   Moderate costs for UPS batteries
          iii. Moderate space requirements for UPS batteries
          iv.   Cost-effective solution while minimizing necessity for colocation/rollover facilities (though this does not eliminate that need)
     b. Disadvantages
          i.   More expensive UPS costs than first approach
          ii.   More space required than first approach
          iii. Does not allow enough time for on-site engineering team to troubleshoot if cascade failures (that cannot be predicted by automated systems) occur.

3. Have capacity >30 minutes (high capacity levels)
     a. Advantages
          i.   Allows Tier-IV performance of electric plant
          ii.   Most robust design minimizes need for colocation/rollover facilities
          iii. Allows effective emergency response by on-site engineers in the event of catastrophic failure.

    b. Disadvantages
        i. Highest UPS costs of any approach
        ii. Most UPS space required of any approach

Within the industry, "conventional wisdom" says it is simply <u>impossible</u> for on-site engineers to adequately respond within 15-30 minutes of a catastrophic failure within a data center. Said wisdom comes from engineering managers, consultants and executives alike. I'll offer a quick story in response:

While conducting annual switchgear and UPS systems testing a data center, some newly-installed sensors within the switchgear were improperly installed by a local electrical contractor. The sensors were improperly insulated. The insulation degraded in a short period of time, resulting in a phase-to-phase short inside the 4160-volt switchgear. The resulting explosion- inside the utility breakers, on the same electrical busses that the backup generators would use- was catastrophic.

The utility breakers tripped open during the over-current condition as designed. The generators- as intended- immediately started. After coming up to speed and proper voltage, the generator breakers closed, re-powering the destroyed section of electric plant, causing a second explosion <u>and</u> subsequent tripping of the generator breakers and their emergency shutdown.

At that point, cooling systems had failed, and IT loads were carried exclusively by UPS batteries, with an estimated 30 minutes of capacity before reliability could no longer be assured.

We were able to quickly determine what had happened. I ordered my team to immediately <u>find</u> a way to get utility power back to the plant, I didn't care how. I ran out, to verify that the UPS systems were carrying the load, as well as call the site manager to inform him of the situation.

Another five minutes were needed to verify the UPS systems were carrying the load as expected. 20 minutes after that, we had devised a means to isolate the damaged section, acted on it, and restored utility power to the facility. Total elapsed time was 32 minutes; we were sweating whether the batteries would last long enough for us to restore power, but we squeaked through. There were no interruptions to IT operations or infrastructure. The IT clients never knew it even happened.

As you can imagine, my *personal* experience contradicts "conventional wisdom," and that a properly trained engineering crew <u>can</u> respond in an effective manner to a site emergency. But such abilities require superb levels of technical training, documentation and on-hands experience with the equipment. IF you have those other assets, the extended UPS battery capacities can be a viable alternative to building vastly more expensive colocation/rollover facilities, while reducing total operational costs to the department.

10 COOLING SYSTEMS IN THE DATA CENTER

The typical house has a 100-amp, 220-volt feed, a normal home consumes- at most- 22 kilowatts of power, and only then for short periods of time. Smaller enterprise data centers typically consume anywhere from 1 to 3 megawatts of power, while larger ones can consume 25 mW or more, on a 24x7x365 basis. The power consumed is given off as waste heat, which much be removed quickly and efficiently. Thus, the cooling system is the opposite side of the coin from the electric plant, and *just as important* in both design and considerations.

Luckily, there are only a few basic approaches to cooling the data center, simplifying the discussion of Data Center Heating, Ventilation and Air Conditioning (HVAC). Those approaches are Direct Expansion Air Conditioning (DX), Chilled Water Air Conditioning, Evaporative Cooling and Economization. While there are myriad technical details that can affect the affect a particular approach, we'll discuss each approach in general terms, and go over the significant strengths and weaknesses of each; one will definitely stand out as having the advantage for your needs.

Air-cooled DX cooling is similar to what you would see in a house or a car, but much larger: refrigerant is pumped to a condenser on the roof (or some other area) outside the building, where ambient air will cool the refrigerant. Fans are used on the roof to boost air flow over the evaporators. The refrigerant returns and is pumped to the air conditioner units in the data center. In fact, these are typically called "Computer Room

Air Conditioners (CRAC) units.

CRAC units have some distinct advantages:
1. Can be installed relatively easily, enhancing operational flexibility.
2. Each unit is a completely isolated system; a leak in one unit affects only that unit, and no others.
3. Small units do not require dedicated engineering room, or require significant disruption to the IT production spaces if repairs need to be made
4. Do not use water, eliminating the risk of leakage underneath the computer-room raised floor reducing electrical shock risk and operational risks to IT equipment.
5. On loss of utility water (such as during an earthquake or other natural disaster) cooling is unaffected.

CRAC disadvantages:
1. Air cooling is not as efficient as water cooling; thus, CRAC units consume more power to operate.
2. CRAC units typically require more maintenance efforts than water-cooled air handlers, due to the increase complexity of CRAC units.

When using water-cooling, there are two different approaches, chillers and evaporative coolers.

CHILLER SYSTEMS

Chillers use water cooled in water towers, where evaporation of some water as it sprays from the top of the tower to the bottom cools the rest of the water. It is then pumped back to the chiller, where through a heat-exchanger, it cools returning refrigerant, which then cools the chill-water which goes out to air handlers on the data center floor. The Computer Room Air Handlers (CRAHs) are therefore different from the CRAC units, though they look similar from the outside.

Chiller and CRAH advantages:
1. Water is a very efficient cooling medium; chillers and CRAHs require less power overall to cool a load, than CRAC units.
2. Chillers/CRAHs require less demanding maintenance than CRAC units; total cost of ownership is significantly lower.

Chiller/CRAH approach disadvantages:

1. High initial cost of installation.
2. High level of difficulty to install additional cooling units as needs change over time.
3. Increased maintenance tasks associated with cooling towers; cleaning, water chemistry, filter media replacement.
4. Increased cost of operation, due to high usage of water.
5. Risk of loss of cooling, should utility water be lost (storage tanks can extend time cooling system can run before water must be restored).
6. Dedicated engineering space required for chillers and control systems- chillers have a large footprint!
7. At partial loads, chillers don't function efficiently
   a. As a practical matter, this means you can't simply install two very large chillers, of say, 600 tons capacity each, and then have 40-60 tons of air handlers for them to cool, with the expectation that you'll add more air handlers later as loads demand.
   b. The chillers require heat loads of at least 30% of their capacity to run, or the chillers will cause short-cycling of the chiller, resulting in possible damage to the compressor impeller and/or bearings.
   c. Using the 600-ton chillers in the example above, if a single chiller is run, it must therefore have 180 tons of load to operate properly, which translates to a minimum heat load of ~633 kW of heat load.
   d. If you don't have that much heat load in your building to start with, you'll have to use smaller chillers to start with, and then install larger ones as demand grows (=more footprint).

EVAPORATIVE COOLING SYSTEMS

Recently, there have been a few manufacturers selling evaporative cooling systems. In effect, these are giant swamp-coolers. The ones I've seen installed are the size of a metro bus- VERY big! They rely on the evaporation of water to cool the air. The highly damp air is then run through moisture separators, and then blown into the data center area.

Advantages:
1. As with water-cooled chillers, water evaporation is highly efficient, resulting in reduced energy consumption to cool a given load.

2. Can switch to internal air-cooled DX backup systems on loss of utility water supply.

Disadvantages:

1. Initial cost of installation appears lower than with chiller plants; however, to take full advantage of evaporative cooling systems, the air flowpaths must be designed to accommodate their use. In effect, they require a dedicated architectural approach to implement.
2. Requires a "whole-room" cooling approach; cannot be effectively utilized in a raised-floor environment (with cooling air coming from underneath).
3. Requires significant maintenance of cooling water systems; chemistry, cleaning of filter media and moisture separators, filters.
4. Risk of mold or other contaminants due to moist air being brought into data center.
5. Increased operational costs due to high consumption of water.
6. Decreased efficiency on days with high ambient humidity, requiring secondary air-cooled DX cooling systems as backup.

A final aspect that data centers can use to maximize cooling efficiency is the use of economizers. This approach, in a nutshell, is to bring in outside air when it is sufficiently cool, and feed it into the data center to supplement the cooling system, if not provide all cooling during periods when local weather is cold enough to warrant it. Of course, this approach requires the data center be in an area where cold air is available, and such air will need to be filtered to remove potential contaminants (such as dust). When properly employed, outside-air economization can result in significant operational savings, due to reduced electrical loads required to cool the facility.

CONCLUSIONS

If a data center is being built where the refresh rate of IT equipment is slow, and anticipated load growth is relatively modest, chiller plants can be very cost-effective and reliable systems for cooling a facility. Their robust nature and reduced maintenance make them a very solid foundation for cooling a large facility. However, the inherent lack of flexibility when tying into a chilled-water system for additional air handlers, and the risks of water leakage under the floor whenever such work is done, makes the use of chillers less than ideal in a data center where flexibility is demanded.

For smaller facilities (<3 mW of power) air-cooled DX units would be ideal, in that they can be brought in on an as-needed basis, eliminating stranded-capacity issues and enhancing agility in the data center environment. As well, eliminating the risks associated with a loss of utility water is significant.

The recent introduction of evaporative cooling systems (swamp coolers) into data centers, is, in my opinion, a poor idea. Between the need for extensive rework of the ventilation systems (using a whole-room cooling approach), insufficient controls to direct air where it is needed, and the environmental problems associated with blowing damp outside air into a data center, all strike me as an inferior option.

A final consideration that must taken into account, is the amount of water the local government will allow your facility to consume. In one plant I managed, a small adjustment- lowering the temperature setpoint by 2°F to accommodate IT requests, increased water use enough (on a 200,000 square-foot data center) to be deemed a "gross water waster" by the local water utility company, tripling the monthly water bill! If water costs are minimal in the area where your data center is, than a water-based cooling system is certainly an option. However, if water is an expensive commodity where your facility is, that should be taken into account for total cost of ownership calculations.

As an example of what <u>not</u> to do, look no further than the NSA data center in Utah. The facility is built in an arid location; water is a precious commodity there. And yet the facility consumes 6 *million* gallons of water per month, with an expectation that the facility will (when at full capacity) consume 1.7 million gallons of water *per day*. If water becomes even more scarce- that is, either more expensive and/or simply not available due to drought or political considerations, the facility is effectively finished as a viable critical facility. No water = no heat removal = no IT operations. WHY did this happen? Because the construction managers failed to consider this significant issue as a potential weakness. Smart managers <u>must</u> question all underlying assumptions, confirming their continued validity. If that confirmation cannot be established to a very high degree of confidence, then contingency plans need to be established, or alternative systems chosen.

COOLING CALCULATIONS FOR AIR HANDLERS:
As with most things, there are easy and hard ways to calculate what the minimum is, for a given engineering problem.

When it comes to determining your needs for cooling, the complicated approach is to use Computer Fluid Dynamics, or CFD. CFD is computerized modeling software to simulate air-flow patterns given a detailed floor plan, locations of cooling units (CRAHs or CRACs), temperatures and air-flow rates of the given units. This is very attractive because it produces very attractive color graphs which demonstrate where hot-spots are, and can also be used to simulate the failure or one or more cooling units within the IT environment, to predict what the results will be. If you're in sales, and need a highly effective presentation tool, CFD software is second-to-none.

The simpler- and FAR cheaper approach- is to use a calculator and some common-sense. You don't get the pretty graphs to show executives, but then again, you don't need to spend obscene amounts of money to get the necessary results, either. As you can probably guess, I'm a big believer in the second method, because the data and calculations are mathematically simple, easily reproduced and not subject to political considerations.

The math goes like this: if we assume the IT equipment will heat up the incoming air by 20 degrees Fahrenheit (commonly referred to as "delta-T" or ΔT), 137 cubic feet per minute of airflow will be required to cool 1 kilowatt of heat.

So if you have a data room where you have IT equipment drawing 200kW of electrical load, you will need (200 * 137) = 27,400 CFM of air flowing through your room, consistently. But, you're no doubt thinking, CRACs, CRAHs and chillers are all rated in "tons," not CFM or kW, so what's going on??? Without getting into the minutia of how cooling is rated in tons (because it really doesn't matter), it's enough to say that 1 "ton" of cooling equals 3.52 kilowatts of heat. That is, exactly 1 ton of cooling will remove 3.52kW of heat load.

Now, in our example, we had a room with 200 kW of heat load. How do we translate this into useful information? A typical Liebert-brand 20-ton CRAC unit will provide (20*3.52) = 70.4 kW of cooling capability. 200/70.4 = 2.84 or 3 CRAC units, running at 95% of capacity, to maintain the environment.

This, however, is NOT all-inclusive. Remember I mentioned verifying assumptions, and there are two underlying assumptions built into this calculation.

The first is that the air is directly exactly where it needs to be, with 100% efficiency, with no "leakage" (which we'll get into later), whatsoever. Since 100% efficiency is *never* attained, let's assume that your IT room is well-sealed and air-tiles are located exactly where they need to be- but that there are still opportunities for improvement, such as hot-row/cold-row containment, better cable-sealing, better-directed airflow measures, and so on. Thus, we can guess at an efficiency (for the sake of keeping this simple) of 80%. We can therefore divide the load by .8, and come up with an *effective* load of 200/.8 = 250 kW of cooling capacity required to keep your IT room cool. This would imply you'll need 250/70.4 = 3.55 or four CRAH cooling units to keep the room cool. See what's happening here?

The second assumption fails to account for electrically generated heat by the cooling units themselves, which will contribute to the heat load within the IT room. 20-ton Liebert CRAH units (which we'll assume are being used with a chiller-plant in this example), typically consume about 26 kW of power, in that *same* IT room. Thus, we add (26 x 4) = 104 kW to the total heat-load in the IT room, which is now 354. Divide that by 70.4 = 5.02 CRAH units to cool the total load. However, because you now have five CRAH units instead of four, your heat load will be 354 plus another 26 kW of heat (for the additional CRAH unit), resulting in 380 kW of heat, a requirement for *six* CRAH units, with a final electrical load of 406 kW total electrical load, to cool an IT load of 200 kW.

The calculations I've laid out demonstrate how to establish the *minimum requirements* for your cooling system. You'll need to factor in periodic maintenance of the cooling units (where they are out of service), and the occasional failure. What happens when you must shut down a cooling unit for maintenance, while another is already offline because of a more serious mechanical failure?

Early on, we quickly discussed "Power Utilization Efficiency" (PUE), and that it is measured by dividing total power consumed by IT load. We also discussed how vendors love to brag about how low they can get the PUE, to be more "green." This quick example easily demonstrates why getting a PUE =<2 becomes very, <u>very</u> difficult, without risking your equipment due to either failing to include redundancy or having unreasonable efficiency expectations (remember, you still have to maintain the equipment while your data center is continuously running). You CAN build a facility with a PUE of 1.6 or even 1.3... The problem is that such results can only come by sacrificing *something*:   reliability, redundancy, whatever.

COOLING CALCULATIONS FOR INDIVIDUAL IT RACKS

Now that we've established the minimum requirements for air handlers in a raised-floor environment, we need to go further down, to the airflow requirements for the individual IT racks that house the servers and storage arrays.

A simple thumb-rule to follow is that the maximum practical load per rack in a raised-floor environment is 3 kilowatts per rack. This assumes that the under-floor plenum is at least 18 to 24 inches deep, there are air handlers around the perimeter of the floor, and you have a ceiling on 8-9 feet. Exceeding this average, you will start to experience hot-spots in your IT racks. The 3kW average is also based on the fact that typical perforated floor tiles can allow up to 300 CFM of airflow to the inlet of IT racks; having two of them in front of a rack yields ~600 CFM. Recall that it take 137 CFM to cool 1 kW of power; 600/137 = 4.4 kW of heat removal, assuming 100% efficiency. The thumb-rule therefore implies an *actual* typical efficiency of 3/ (4.4) =68.1% or ~70%, which honestly isn't too bad.

But suppose you have a situation, where you have a rack or two, where heat load well-exceeds this average? This happened to me, when the client brought in two heavily loaded racks with "Blade" servers; the loads were 9 and 11 kW.

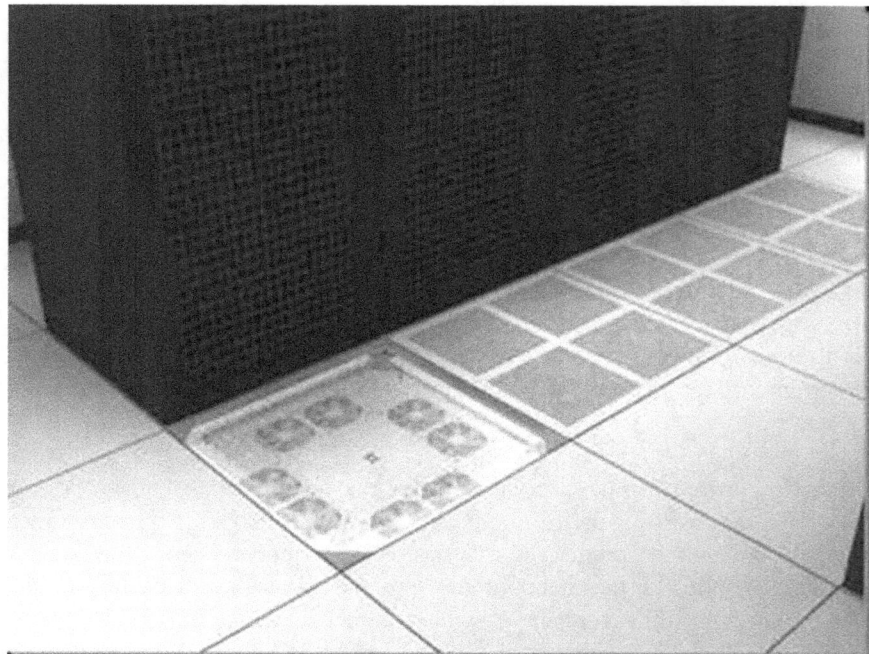

In such a situation, simple perforated tiles will <u>not</u> be able to supply sufficient cooling air, no matter what you try. Instead, active cooling must be used. In that situation, we used "fan tiles." These are tiles that have fans built into them, with louvers to direct the air to a specific location, and temperature sensors to automatically adjust air flow to the rack and temperatures changed with heat loads. This simple solution- while not economically feasible for large-scale operations- can be quite effective for the occasional surprise brought into your facility.

Going back to our original calculations for minimum requirements for air handlers, we used a baseline for efficiency of 80%. You may be there, or even better, but there are some surprisingly simple and cheap ways, to improve the overall performance.

Fair warning: while these suggestions seem overly simplistic, most facilities that I have seen are lacking in some- or all- of them. The implementation of these measures can have dramatic results, reducing hot-spots, extending IT equipment reliability and saving money in electricity costs.

1. Maximize the flow of cold air to the perforated floor tiles in front of the IT racks:

    a. Minimize underfloor obstructions which would disrupt the flow of air from the cooling units to the needed locations

    b. Always remove unneeded or abandoned cables as a standard house-keeping practice

    c. Seal all cabling holes cut in floor tiles, so air is directed to proper air inlets instead of the bottom of the cabinets (or equipment that does not require cooling).

    d. Put perforated tiles only in front of IT equipment racks, and open dampers to deliver sufficient airflow to IT racks, but not *excessive* airflow. 600 CFM of air to a load requiring 300 CFM, means the extra air is not delivered to where it *is* needed.

2. Hot-aisle/cold-aisle containment

    a. IT racks should be organized in long rows, with

        i. Air inlets all on the same (cold) side

        ii. Air exhausts on the same (hot) side

        iii. Ideally, hot and cold sides should be isolated from each other

    b. If your IT racks are not currently arranged in this way, install some new empty racks in an appropriate configuration, and as you refresh your IT equipment, migrate them to the new racks, until the old ones are empty and can be removed.

    c. Do not allow the exhaust heat from one IT rack to flow to the cold-air inlet of another rack.

Cold aisle containment

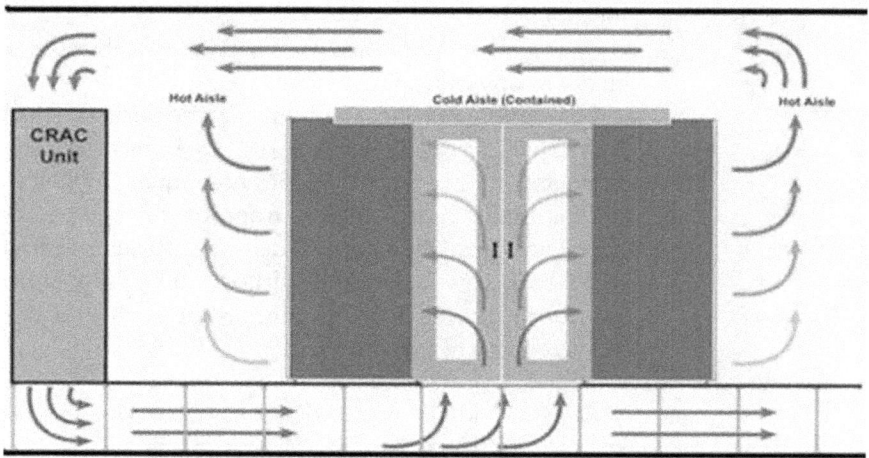

Picture from www.42U.com

In this example of containment, the cold aisle is completely isolated, with the escaping heat returning to the top of the CRAC unit freely. Notice no cold air leakage under the floor, or from the racks. Cold air is drawn in by the internal fans of the IT equipment as needed.

The only disadvantage of enclosing the cold aisle in this way, is possibly running afoul of local fire codes, where such containment may interfere with fire-suppression/sprinkler systems. Check carefully with your local fire inspector, on what options are available.

3. Minimize leakage above the raised floor:
   a. Seal all perforations in floors and walls where cooling air might escape from, so that the cooling air will be sent only to IT loads.
   b. Try to eliminate any paths for hot-aisle air to recirculate back to the cold aisle.
   c. Eliminate gaps under or in the racks where cold air can leak through without cooling actual IT equipment (blanking plates).

A case study, titled "Green IT And Data Centers," is included at the back of this book (Appendix III), so you can see what the actual results were for the facility involved. The total cost outlay for our efforts were $8000, and energy savings were ~$300,000 annually.

OTHER COOLING CONSIDERATIONS

As mentioned before, the average heat load in a computer rack using a raised-floor system is limited to 3 kW, as an average. Once the loading exceeds this level, additional or alternative cooling approaches need to be considered.

Between roughly 3-10 kW of heat load, "row" cooling is a viable approach; cooling units are placed on the ends of the IT rows of racks (referred to as "pods"). Hot air is captured on the ends of the sealed hot aisles, directed through the cooling units that are at the end of the rows, and the cooled air is directly fed back to the sealed cold aisles. There biggest potential disadvantage to this approach is fire safety; a gaseous fire suppression system and/or sprinklers would necessarily have barriers to their effective deployment during a fire; check with your local fire marshal to determine what is permissible.

Other disadvantages of row-based cooling is that redundancy is required at each suite of racks; unlike the raised-floor system where one or two air handlers can provide redundancy for an entire room, the row-based cooling units must have redundancy built into each pod, increasing the capital cost of this approach. As well, the fact that the cooling units are in the very aisles they are cooling, requires that any maintenance be performed in the aisle, where access may be limited. Finally, the cooling capacity of the row cooling cannot be shared between pods; each is isolated from the next.

At heat loads exceeding 10 kW of average heat, the safest approach is through rack-cooling, which is, in effect, the installation of large radiators on the inlet of those racks, allowing chilled water to cool the air dramatically, before it enters the rack. (Alternatively, the radiators could be placed on the exhaust side of the IT racks, but there can be losses in efficiency using this approach.) Rack-cooling eliminates the various airflow leakage and bypass issues of either the raised floor or row-cooling approaches, but requires the radiators be fed from a cooling system, leading to potential risks of coolant leakage in the data center. The radiators must have connection points and isolation valves in the event of coolant leakage.

The radiators also make access to the IT equipment problematic- the radiator must be swung out of the way (much like a refrigerator door), thus removing the cooling from IT equipment that you will want access to. This therefore requires a method of redundancy for each radiator (doubling the amount of connection points) and driving up the initial installation costs dramatically.

FINAL THOUGHTS AND A REAL-WORLD EXAMPLE

As you have no doubt gathered, the cooling system chosen for a data center must be chosen as carefully as the electrical system; mistakes on either side will result in very expensive mistakes that can easily destroy your budget or cause systemic failure of your facility.

A real-world example is a data center that I was involved with from conception until final commissioning. The IT managers anticipated a maximum heat load of 15 kW per IT rack, and each pod would have 160 racks.

A whole-room cooling approach was used, where evaporative air handlers installed on the roof would provide cooling, while still being very energy-efficient. The IT pods were set up that the air would dump from the overhead into the sealed rooms, and the air-flow would distribute to where it was needed, drawn by the IT rack fans.

With the anticipated 2400 kW of heat per pod, assuming the ΔT for the air is 20°F, we now have our required airflow: 333,000 cubic feet per minute of airflow. A massive duct for air inlet would be required; if we used a hole 8 feet tall by 16 feet long, we would result in an air velocity of ~32 miles per hour, assuming 100% efficiency, but at a more realistic 90% efficiency, we're now up to a huge hole blowing down air at 35mph, and we will need an equally massive air return (if not more so, since the hot air has expanded). In essence, the design called for a wind-tunnel to be built, which would be impossible for IT or maintenance personnel to work in. It would also be a nightmare for controlling the air; that is, the hotter racks get more air than the less-loaded racks.

When I provided this information to the construction firm performing the build, the information was summarily dismissed; I was not a Professional Engineer (PE) nor do I possess a formal engineering degree.

As construction progressed, the limitations of the ducting system, air control approaches and finally the physical limitations of working within the IT pods became apparent, and maximum allowable heat loads in each rack were subsequently downgraded multiple times. At the time of commissioning, the specifications of the facility within the commission plans said the system would be tested to be 5 kW of load and 3 kW of load, within the same document. Ultimately the facility was commissioned to a 3 kW threshold, because final testing revealed that higher load densities could not be reliably cooled without hotspots forming. Now, after the facility has been in operation for an extended period of time, control issues with regard to airflow continue to plague the facility.

Lessons to take away from this example: basic calculations can be invaluable to determine whether something is even feasible. If the calculations say "no," then view the fancy solutions that vendors are trying to sell you, with increased skepticism. Another lesson from this, is that professional engineers do- quite often- make mistakes. Sometimes the mistake is small, sometimes it is large and obvious. Don't think that the PE certificate magically guarantees you'll be satisfied with their approach, or that they can somehow defy the laws of physics (as the above example demonstrates abundantly). Finally, trust your teams to give you a true analysis of what is being proposed; their experience levels will provide a "gut-check" that is priceless, and can save the firm a LOT of money.

11 SYSTEMS: PUTTING IT ALL TOGETHER

So now that we've gone through the various pieces of a mission-critical facility, we can use a past example to show how everything coalesces into a cohesive support system for a critical facility.

The project we'll use involves the replacement of power and cooling systems in an existing facility.

Physical location: The building is on a large campus with multiple buildings, and the perimeter is enclosed cyclone-fencing topped with razor-wire facing outward. Access gates to the campus are manned with security guards, who were frequently tested to make sure they were vigilant. The campus itself is ~2 miles from the nearest highway or rail-way line, and 20 miles from the closest airport.

The geographic region is not in a flood plain, but experiences occasionally strong seismic activity.

The building itself is a 75 year-old concrete facility, almost a bunker, where additional office-space was created by building additional wings, of lighter materials, though the bunker itself remained intact behind the less robust offices. The perimeter of the building is surrounded by concrete bollards, sufficiently large and strong to stop a large vehicle like a semi-tractor.

Access to the building required magnetic ID cards, and the access lists are checked frequently (periodicity is shorter for vendors and as subsequently deeper layers of the facility).

ELECTRIC PLANT

The new electric plant uses two feeds from different electrical grids.

The original plans called for manually-operated breakers, with two automatic bus transfer switches (ABTs), as shown:

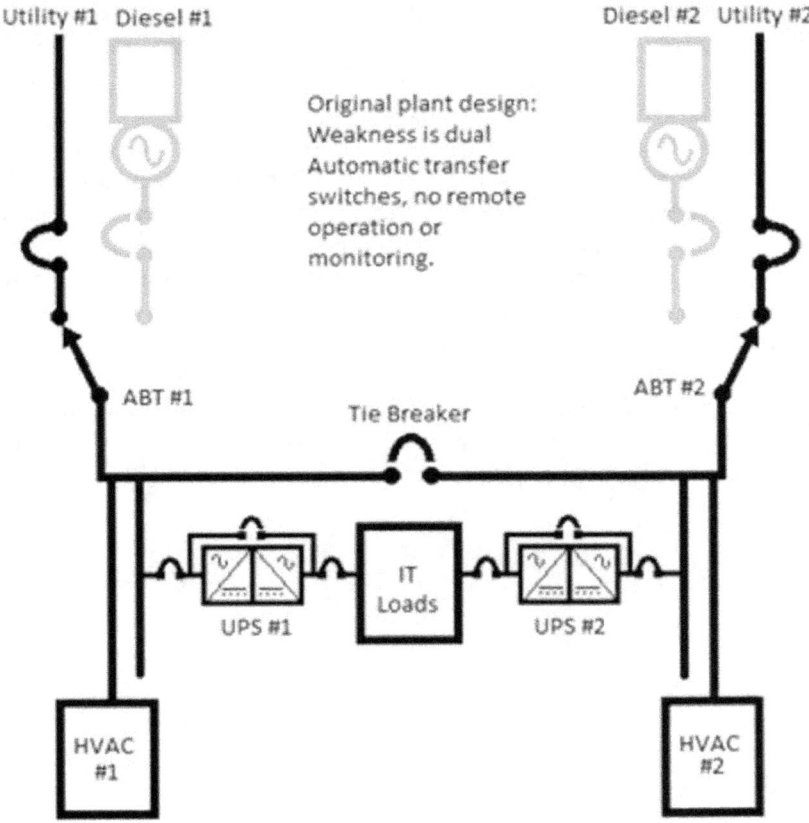

When failure of a utility feeder (#1 for this example) occurs, the utility feed breaker #1 opens, generator #1 starts. When generator is running properly, the transfer switch detects a new power source, and switches to it. Recovery requires manually reclosing the utility breaker, and then performing a manual shutdown of the generator.

There were no on-staff personnel properly trained to operate the electric plant in normal or emergency conditions.

The UPS systems chosen were manufactured by Liebert, with each UPS able to carry all current and planned-for-the-future loads. The Liebert UPS units- though not the very best in energy efficiency- are robust, long-lasting and used widely in the critical facilities industry. As to battery capacity, the desire was to have 90 minutes using AGM batteries. The excess capacity required additional space within the limited floor of the building, and that much capacity seemed a bit of overkill; however, the client was interested in having the longest amount of battery capacity possible for the facility, regardless of cost or logistical issues.

The above design had several inherent weaknesses. First was the fact that there were no properly trained personnel to operate the plant, to recover from a loss of utility power. Some staff members had been operating the system in the past, but were not fully trained, and unaware of safety protocols and operating rules. Another weakness is that the tie breaker was manually operated, and could not automatically operate for certain conditions such as a failure of the ABT to operate properly. Finally, most ABTs have a failure mode that ties together two independent sources of AC power in an uncontrolled fashion, resulting in catastrophic failure (usually an explosion). Finally, if the ABT failed to operate during an extended outage, the associated UPS would eventually suffer from a dead battery, and the critical power (from the UPS on that side) would fail. A final flaw is that local utility power can be somewhat unreliable; each time a utility feeder is lost, the generator would automatically start, and could run for several hours before the plant was restored to normal configuration.

An alternative design was proposed:

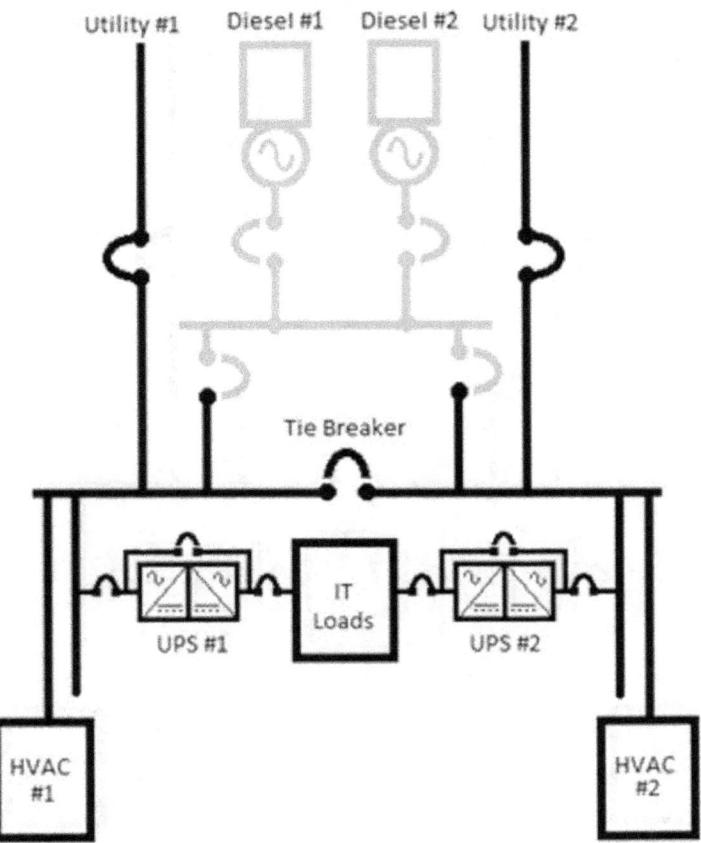

The ABTs would be removed. Utility, generator bus and tie breakers would be operated via Programmable Logic Controller (PLC) a small, simple computer. This would eliminate the need for human intervention on a regular basis; the system would respond automatically to any normally expected event. The removal of the ABTs offered a significant cost savings, as well as saving precious space within the confined location. Finally, the generators would not operate unless both power feeds had been lost, and then either one or both could start and assume the load.

An example of what would happen on the loss of single utility feed is shown in the below:

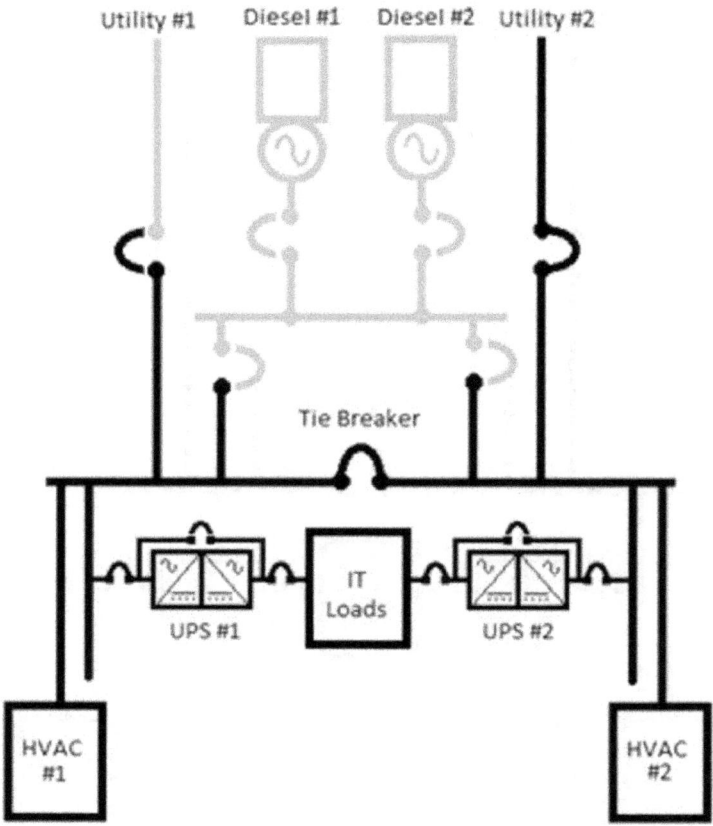

In this example, Utility Feeder #1 power is lost. The PLC automatically opens the utility feeder breaker. After breaker sensors indicate the breaker is indeed in the open position, then the Tie Breaker closes, restoring power to all downstream buses.

There would be a momentary loss of power to HVAC group #1 and UPS #1, but could be limited to ~5 seconds.

On restoration of utility power on feeder #1, the PLC detects power availability, opens the Tie Breaker and recloses utility breaker #1, restoring the facility to a normal configuration. Notice, in this example, the generators never start.

The client was uncertain of the reliability of a PLC-driven system, being unfamiliar with this technology (has been used effectively for >40 years). The compromise solution that was finally installed closely resembles the original design, but with some important improvements:

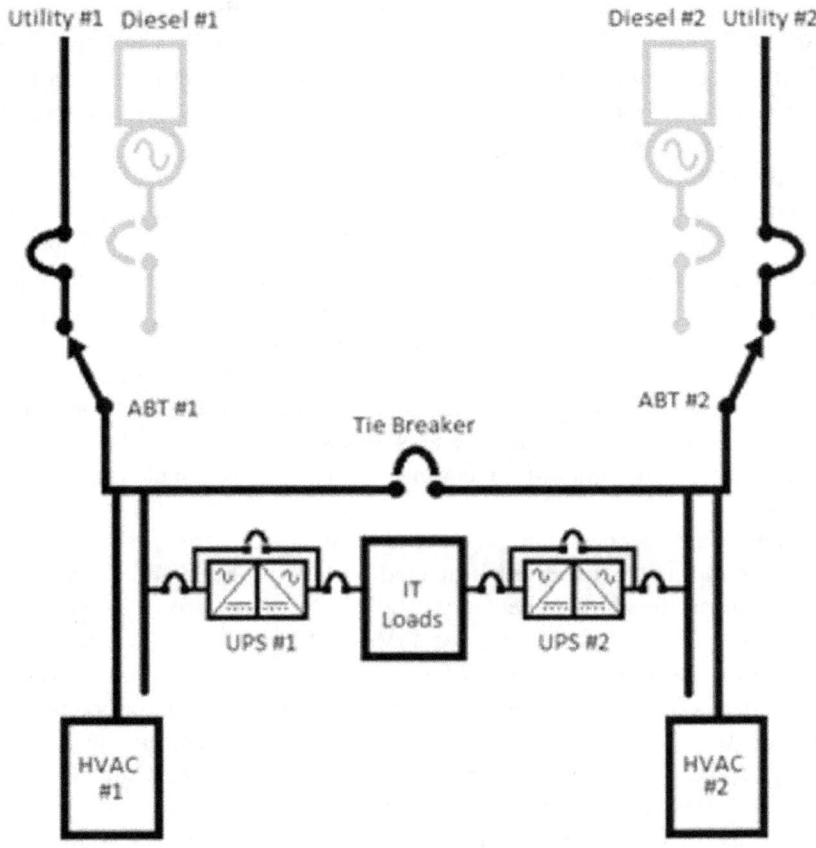

The client insisted ABTs be installed, due to past experiences with them. The Utility and Tie breakers and ABTs were linked to the PLC. On loss of utility power, the PLC would automatically open the associated utility breaker, and close the Tie Breaker, restoring power. If the tie breaker failed to shut, it was "locked-out" on error, and the generator would start. The ABT would then switch to the generator, and restore power. If both utility feeds were lost, both utility breakers would open, the tie breaker would stay open, and both generators would come online, restoring power. If a generator subsequently failed, the associated generator breaker would open, and the PLC would shut the Tie Breaker, restoring power.

As you can see by these simple diagrams, the high-level design of an electric plant can be very simple to understand, and also very reliable. The final design was flawed by the inclusion of the ABTs and inability for the generators to work together in parallel, limiting flexibility and spending extra money that- in my opinion- offered little benefit in comparison to a fully PLC-driven system. On the other hand, the customer was very happy with the final results. After experiencing a variety of fluctuations on their utility feeds which would have been annoying with the original design, the system proved extremely reliable.

MECHANICAL PLANT

The campus that the building was on had multiple water-mains from the local municipal water company, and had no existing worries about loss of cooling water for their critical facility. However, because the facility was quite old, with limited interior volume and no room for chillers, Computer Room Air Conditioners (CRACs) were used. These are air-cooled, direct-expansion air conditioners, basically upsized versions of what you would find in a residential home. This approach allowed the cooling coils to be installed on the roof with auxiliary fans to assist efficiency on hot days. It also minimized the amount of cooling infrastructure installed within the facility.

There were sufficient CRAC units to have 2N cooling. That is, twice as much cooling as the expected heat load within the building. One group (referred to on the above schematics as HVAC #1) was tied to utility power on the #1 feeder, and likewise for HVAC #2. All units would run in normal operation, at some loss of energy efficiency, but if there were a momentary loss of utility power from either feeder, the opposite side offered sufficient cooling (plus thermal inertia) to overcome any decrease in cooling capacity by the momentary loss of one group of CRAC units.

SYSTEMS CONTROL

As mentioned previously, the entire electric plant control system was based around a programmable logic controller, or PLC, similar to the picture below.

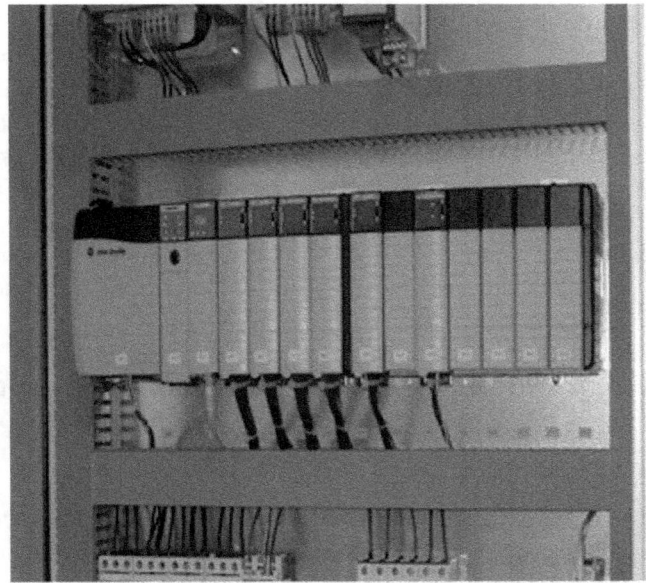

PLCs are a hardened computers, designed to operate in hostile environments (temperature/vibration/humidity). Programming is accomplished via ladder-logic software, using simple and intuitive programming interfaces. While PLCs don't have extensive memory or processor speeds like the typical home computer, the operational programs needed for operating an electric plant (or factory, for that matter) are comparatively small and simple. Their robust design makes them ideal for data center use. There are several excellent brands of PLC, including Siemens, Allen-Bradley, ABB, Mitsubishi and Schneider.

The ease of programmability means they can be adapted to changing conditions or priorities within the data center environment, as well as adapting to a variety of potential casualty situations. The key, of course, is those possible casualties (such as loss of a utility feeder, or even a *series* of faults) are programmed into the PLC. If a situation develops that nobody foresaw, then the PLC may not react, or react sufficiently, to successfully respond to cascading events. IF cascading events can be estimated (even if highly unlikely) and accounted for, PLCs *can be* superb.

Another positive aspect of PLC controls, is that they can operate very quickly, far faster than a human can. The response time for events can sometimes be crucial, and PLCs can hold the edge over their human counterparts. If responding to a cascading event, engineers must take time to understand the problem, devise a solution or way to slow/stop the cascade and gain stability, and then put the solution into effect. Again, if the PLC is properly configured with solutions for potential cascades, the

PLC can be several orders of magnitude faster.

This doesn't mean that PLCs are perfect; they require careful installation, absolutely *top-notch* programming, and rigorous testing during the commissioning phase, after installation and programming are complete. That means testing for not only individual faults and events, but simulating and testing for potential cascade failures. This is not easy, and it's especially uncomfortable, because of the work and pressure you're under. Keep in mind, such upgrades rarely occur in an empty facility; they're usually done in an operational facility, where the clients are *not* happy with testing on their critical support systems.

The last issue that PLCs have, is that many engineers of the "old school" variety are uncomfortable with systems where they cannot examine internal components. In other words, older engineers are nervous about "black box" technology, which they don't understand.

RESULTS

The overall result of the electric plant design is that the electric plant could absorb 3 out of any four potentially possible faults (including failure of one or both generators plus a utility feeder, or both utility feeders plus a loss of either generator) in any sequence, and still maintain service to the critical IT equipment. More importantly, if conditions changed in such a way that a cascade failure started to occur from one to multiple faults, the PLC would detect changing conditions, and modify the electric plant configuration to accommodate those conditions, *still* maintaining plant operations. This is what is referred to as "self-healing" capabilities, a key point of a Tier-IV facility.

After commissioning of the facility was complete, the entire geographic area experienced multiple power outages, where several utility grids were going offline, coming back, and then failing again, at different times and with no discernible pattern. As this was occurring, I called the client to see how they were doing. They had no idea that anything at all had happened, because the system literally adapted to each situation flawlessly, and transparently.

Due to the way the utility power is fed to the CRAC units, loss of a utility feeder would necessarily cause a power interruption in those CRAC units for a matter of a few seconds while the PLC detected the fault, opened the appropriate utility breaker and closed the tie breaker (or if both utility feeders were lost, while generators start- an operation of $<= 30$ seconds). It would then take 90-120 seconds for those CRAC units to restart, and begin providing cooling to the facility.

In a facility where the power load is low (average is <3 kW per rack), the described interruption in cooling typically is not of sufficient duration to cause overheating of the IT equipment. However, Tier-IV facilities are specified as having "continuous cooling." That is, even in the event of a utility outage as described, the cooling system would suffer no interruption of cooling capabilities. In medium (3-10 kW per rack) and highly loaded racks (>10 kW per rack) the demands are far less forgiving, and such an interruption could cause serious problems.

Continuous cooling requires significant changes in cooling system power feeds, including UPS power to oil pumps, rotary UPS systems or other solutions to preclude a utility failure from affecting the cooling system. Continuous cooling **cannot** be accomplished using solid-state UPS systems, because the starting power drawn by large motors during startup (such as chillers or multiple CRAC units) can be up to five times normal load, which would cause solid-state UPS systems to- at best- fail on overload and shut down, or- worst case- fail catastrophically (explosions, fire, or other very bad stuff).

Recall that a true Tier-IV facility requires continuous cooling to be given this prestigious award. As I already pointed out, there are times when that is necessary. But for a facility that expects relatively low power density, I have come to the conclusion that continuous cooling is often an unneeded expense. IF the cooling system has sufficient redundancies and the ability for concurrent maintenance, and this is matched with an electrical design that has multiple redundant paths and self-healing capabilities, the result can be an extremely robust system while still being economical.

12 PLANNING AND COMMUNICATIONS

Strategic Management of Mission Critical Facilities

Up until now, the information in this book has dealt with day-to-day operational and considerations. However, a key aspect of managing data centers is strategic management; looking into the future months and years ahead of time, and making appropriate plans and decisions which will assure long-term viability of the facility. In a data center or portfolio of data centers, we'll define the different time-frames by saying the customer needs within the short-term are operational management issues (those that occur on a day-to-day basis), tactical management is of the medium term (<1 year out) and strategic goals and objectives are long term (>1 year out).

For the construction of a new data center, upgrades or significant alterations to an existing facility, strategic goal-setting involves several guesses; how much floor space is going to be required? Can the goals be accommodated using the current raised-floor (or above ground-level) or will there need to be additions? Can the changes be installed on an existing concrete slab that we already have? If it's a raised floor, what is the maximum anticipated floor load in terms of pounds per square foot, that the floor will be expected to support? If it's a raised floor, how much power per computer rack (power density)? What is the *maximum* amount of critical power that the facility is expected to support? How much load is anticipated when the facility first goes "live?"

When you're building a data center- or simply renovating an older facility- these questions obviously need to be asked. They're as basic as an aerodynamics engineer building an airplane; how much will it have to carry, how fast, how many people, and what are the maneuvering expectations?

Theory: At a company-wide level, the textbooks tell us that strategic management consists of the following management circle, which we've all seen before in some fashion or other:

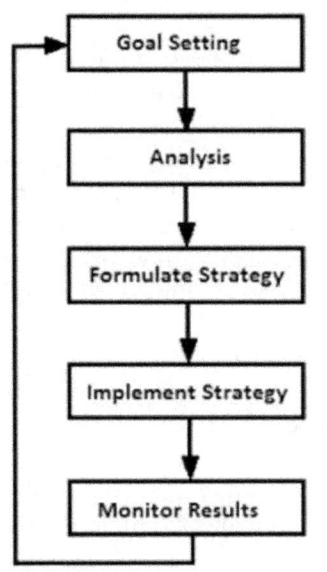

The typical cycle of strategic management starts out with the following steps:

<u>Setting goals and objectives for the facility</u>. "Where are we going, and how do we get there?" This is where the data-center build-out is conceptualized, with maximum power loading, cooling strategies, fire and life-safety systems are considered and floor layouts are planned.

Comparing the long-term goals and objectives to the current market conditions. "Where are we now?"

We look at exterior forces such as market conditions, anticipated technological changes, external cost and anticipated changes, and even considerations that could change how your business operates. We would look at internal forces such as anticipated revenues and internal costs versus potentially outsourcing costs, appetite for risk (current Tier level of your mission-critical facility versus future needs for reliability that your business model may require), and the current amount of flexibility and agility versus what future requirements.

Finally, consideration must be taken for not only how the management of the facility will be conducted, but also the work environment for your support teams.

<u>Data collection, analysis and interpretation</u>. The data will necessarily be of both quantitative and qualitative types, *with the latter being more significant to strategic planning* due to its perspicacity. For example, gathering information on the effectiveness of communications between clients and the service providers (in this case IT departments to critical facilities staff) should be carefully assessed; is there an effective means to communicated changes and receive feedback to those proposed changes? Do the clients and service providers talk directly to each other? The data collection, to be meaningful, requires internal leaders to assist in the collection and provide meaningful

feedback; are the goals and objectives realistic? Can the goals and objectives be met within existing constraints (such as power, cooling and floor space)? If not, what changes will have to be made to existing (or future) infrastructure to assure client needs are met?

Formulate the strategy, through projects and initiatives. Now that you know where you are, and where you want to *be,* the hard work really begins; figuring out how to make those goals become reality. This may be something as simple as adding a wing onto a building or upgrading your CRAC units to improve cooling of hot racks, or it could include building a completely new facility from the ground up. This part of the strategic plan will also include decisions on how to acquire capital, how to procure contracts, estimation of project staffing (both internal and external) and a variety of other details that are required for success. This is when you'll develop project scopes, deliverables, timelines, budgets, etc.

Implement the strategy, in accordance with the just-finished plans, and

Monitor the results in an on-going basis, and adjust accordingly so that results match expectations.

At this point, the strategic plan is written, with project deliverables, timelines, budgets, scopes of work, etc. Pretty straightforward stuff, right? Uhh... Actually, *no...*

EFFECTIVE COMMUNICATION

It seems so obvious when you think about it, but management of a team, department or division, relies on effective communications. This includes subordinates effectively communicating information about conditions, opportunities, ideas and recommendations, as well as the managers' effective communication of their analyses of the data (or requests for additional information), concerns, decisions and plans for action *It's **all** about effective communication,* and without it, you're in deep trouble. The need for an effective communications model is critical. The effectiveness of your communications model will determine whether your strategic goals are met. Let me repeat that:

The effectiveness of your communications model will determine whether your strategic goals are met.

HOW DOES THIS APPLY TO DATA CENTERS?

While the following description sounds a bit unkind, the following description sums up the majority of my observations from the industry. I'm reasonably certain a majority of others in similar jobs have seen similar results...

The strategic management feedback loop, in my experience, is something that consistently *fails* to function in the real world, because the decision-makers within companies (especially the large ones) make their decisions in a bubble; they fail to consult with people who work with the results of those decisions every day. Sadly those same decision-makers, should they realize they're not making *fully informed* decisions, will turn to outside consultants who may (or may *not*) provide meaningful information. The "demotivational" website www.despair.com summarizes this nicely:

So the decision-makers often make strategically *un*sound decisions, because they're simply not receiving valid input from the building engineers who have the most experience and talent to make those goals become reality. Power and cooling capacities are planned out by degreed engineers and consultants who have little (or no) experience in the practical day-to-day operations of data centers, and *certainly* will not suffer the consequences of a design mistake (short of finding a new gig to consult on).

DATA ANALYSIS: With flawed goals, the incoming data will appear confusing and/or contradictory. The building engineers who will be operating the facility will argue with the design engineers and consultants over whether the new design of mouse-trap is necessarily better than the old one. The old-timers will rely on decades of experience and "seat of the pants" assessments, while the design engineers and consultants will rely on their computer models, predictive programming and 'sell the sizzle' on things like energy efficiency, "green" computing and the like. When all else fails, the design engineers and consultants will appeal to authority (*theirs*, of course) as they trot out their degrees and certifications, ultimately winning the day with the decision-makers. The project will move forward as planned.

IMPLEMENT THE STRATEGY: At this point the designers and consultants are effectively in charge, and build according to their now-approved design. In a quest for marketable sound-bites, reliability and redundancy take a back seat to a few additional percentage points of energy efficiency, and the ability to proclaim the facility LEED "Gold" or "Platinum," while the underlying infrastructure systems that the entire company depends upon are degraded.

In the rush to construct the new facility so that the company can proudly unveil their newest technological wonder, the construction time-line is shortened, forcing the project managers to dash through the commissioning steps, ignore glaring deficiencies and rush to complete the project. The "keys" are handed to the building engineers who to their chagrin are now tasked with making a deficient facility match the unrealistic expectations of the decision-makers- or *else*.

FEEDBACK LOOP: When the building engineers take ownership of the new facility, they quickly determine that that facility is in poor shape; systems don't operate as they should (even according to the less-than-optimal design), safety systems are non-functional or operate inconsistently, energy efficiency doesn't match the promises of the designers, and the overall result is a facility that is unreliable, not sufficiently redundant, and requires a significant amount of capital in repairs and upgrades, to get the facility to match the original promises. The consultants- helpful bunch that they are- are quite happy to come in and suggest corrective actions for all the glaring deficiencies.

The decision-makers are livid; how could this have happened? The feedback loop is broken, because those same decision-makers were the *origin* of the entire failed process, and they are simply unable to comprehend that the responsibility lies with them.

WHAT WENT WRONG???

"What we got here is... failure to communicate." Captain, Road Prison 36. *Cool Hand Luke (1976)*

The failure is that the decision-makers (usually IT executives) don't develop their strategic goals with sufficient information. Such goals must be developed with direct input from those that will be held responsible for operating the facilities after construction is completed; the building engineers and facilities managers.

This is natural byproduct of the Information Technology Infrastructure Library (ITIL) which has been the model for strategic management, planning, communications and feedback of the IT industry since the *1980's*. It is an old system (with a few updates), and is considered the gold-standard for IT strategic and operational management. Unfortunately it has a hidden flaw; it *automatically* assumes that adequate power, cooling, floor-space and water are always available. It therefore removes from consideration the building engineers and facilities managers from the equation. The people most qualified to help develop meaning, *actionable* plans are left out of the loop- by design!

Given the ITIL methodologies being an industry standard for better than three decades, few IT executives know anything *before* ITIL came along, so they simply don't know that they're setting themselves up for failure.

HOW STRATEGIC MANAGEMENT OF DATA CENTERS SHOULD BE

Going back to our original strategic management loop and applying it to the concept of building a new

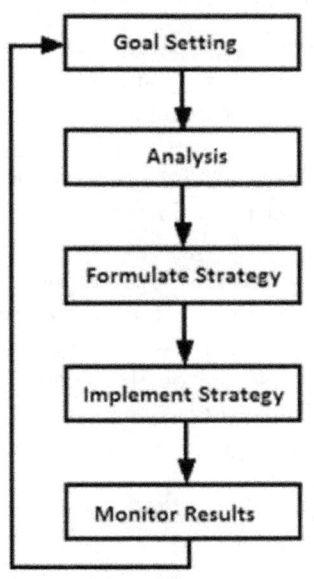

data center, we start by setting realistic, attainable goals and objectives for the facility. "What is the plan for this facility?" What is the realistic data-center size, initial and ultimate goals for power loading, initial and final cooling strategies, fire and life-safety systems? How do we best utilize the floor-space to maximize initial and long-term goals, with maximum flexibility? What is Tier-rating for the facility to be? What is the feedback from the engineers who will be operating the facility, to these goals?

We then analyze the business case for a new construction: do external market conditions and anticipated revenues from this site justify the construction, or does it make sense to rent space in a colocation facility (buy vs. rent)? What is the appetite for risk, at the company? Do you have adequate staffing (and labor budget) to match the Tier rating you desire for long-term sustainability of the facility? Where do you plan to build it, and what are the local geographic risks you will have to mitigate? Local political risks like taxes and crime?

When you've done a detailed analysis, the formulation of strategy must also be true-to-life. The construction project plan must include sufficient contingency money for unexpected bureaucratic snafus, location-specific risks and construction of sufficient infrastructure to make sure the facility has adequate services.

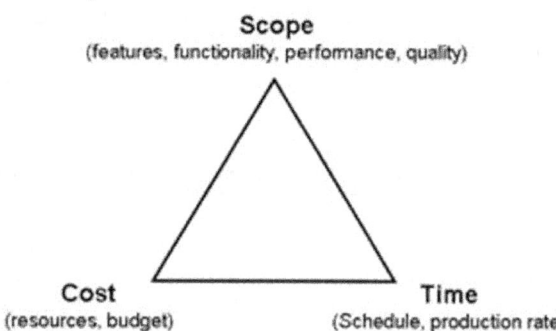

Scope
(features, functionality, performance, quality)

Cost
(resources, budget)

Time
(Schedule, production rate)

The implementation of the strategy is where it seems every data center owner fails miserably; the project management triangle. The owners/executives that develop the goals and strategic plans always assume that the construction companies will meet incredibly short time-lines, delivering a vast scope of work and goodies, all for a very cheap price. The reality- unsurprisingly- is that such promises are *never* fulfilled.

I keep saying this, because it's so important in the mission-critical environment: *be realistic in your plans.* If you had to go in for open-heart surgery, how realistic is it to believe a cardiovascular surgeon who claims he can do a quadruple-bypass on you (or a loved one) for 30% less money out of pocket, have you home by tomorrow night, <u>and</u> offers a guarantee that you'll live at least another ten years afterwards? Would this not strike you- especially if you've done even a smattering of research on the subject- as somewhat dubious? Construction of your data center is the same way; construction companies will promise the *world* to get your business, and after construction is underway (and delays start to pop up) you're already *stuck* with them. At that point, you've no choice but to see it through to the painful end, even if you're 25% past your "deadline."

So let's go through the construction project management aspects of a data center.

SCOPE includes the physical aspects of the data center, including construction, overall footprint of the facility, including data halls (or pods), engineering spaces, offices and storage areas, IT staging areas, loading docks, security offices and so on. The scope should also include the Tier level of the facility (performance of the infrastructure), and include the deep-dive details for the monitoring, automation and control systems.

TIME should include realistic timelines for the construction of the facility, including very specific deliverables. You MUST establish priorities and dependencies, and document these in a project schedule. Typically, if some dependencies are not fully met, a construction company and/or project manager will allow construction to continue in the hope that the dependencies can be met at a later time.

For example, in one construction project utility power was actually fed into the site backwards; phase rotation was reversed, but was not detected for two months because that portion of the testing was delayed. When discovered (by accident) all work had to stop and systems rewired to match the corrected phase rotation, causing weeks of delays.

Another tactic project managers will sometimes take is to throw extra assets at a problem, thinking that by multiplying the work-hours associated to a task, the calendar time associated with that task is inversely shortened: triple the staff, and you'll need 1/3 the time, right? Umm... again, *no*. The flaw in this line of thinking (which happens all the time) is summarized by a client who once told me, "that's like getting nine women together and thinking you can get a baby in one month."

You *must* be realistic in your timelines, and while it is important to be on-time, you must allocate contingency *time* into the schedule just as you would contingency funds. This is the nature of the beast, due to the sophistication and performance expectations of high-performance data centers. Failure to take this into consideration *will* result in disappointment, of that you can be certain.

TIME should also include the sequence of activities, estimated times to complete each activity, estimated resources to complete the activity. From these estimates, a full schedule can be developed.

The activities themselves need to be defined as well as tools and testing to validate that each activity is complete and performs exactly as intended, to maximum design limitations.

COST would seem a self-evident thing, but includes variables like labor rates and other issues that can cause cost variances. Projected costs can be evaluated by comparison to other similar projects, analysis of vendor bids, quality cost-analysis, parametric analysis or determination of rates for resources.

Overall project management is constantly a dance between the cost, time and scope. The project manager can always deliver two out of the three demands, at the expense of the third. Clients always demand all three elements, which is impossible, as a practical matter. So the data center project manager always tries to juggle an impossible situation, resulting in inferior/incomplete commissioning of equipment, cheaper components for infrastructure, questionable reliability, inadequate documentation, shorting of attic-stock (spare parts) when the facility is handed over at "substantial completion," inadequate (or no) training of the operational staff, poor operation of automation and monitoring systems or lacking warranty support after the construction is complete. On the construction efforts I have seen for each mission-critical facility, every example had not just one, but *several* of the above deficiencies.

The reason why this occurs repeatedly is that the project managers are required to meet time, scope and cost constraints, but they are ***never*** required to balance *quality* of results as a constraint like the other three. Thus, quality is merely *assumed* to be delivered as a natural byproduct, despite rushed commissioning, an incomplete punch-list at the end of the construction (such punch-lists are rarely completed before delivery to the client), inadequate documentation and training of the staff and poor customer-service after delivery of the facility.

And I see this over and over, with both small firms and large corporations.

Getting back to our strategic management loop, with unrealistic implementation plans being forced upon the project managers, the results are entirely are predictable for the mission-critical environment; disappointing performance of the facility in terms of inadequate power and cooling delivery, poor reliability and redundancy. However, since executive managers typically underestimate the time required (and make unrealistic promises to clients), they're either forced to ignore the feedback loop which says the facility is not ready for prime-time, or they simply accept the facility, warts and all, and hope that the operations team who will be responsible for running the facility can somehow clean up the mess after the fact. Thus, the feedback loop (currently) is defeated.

It should be noted that there are other costs associated with failing to include a meaning quality metric into the project; higher than anticipated total cost of ownership due to capital repairs to later get the facility to deliver as intended, thereby yielding a reduced return on investment. Those facilities that intend to lease space to 3rd party clients will also find their customers less confident in the reliability and redundancy of the facility, and experience lower profits due to incurred costs when service level agreements (SLAs) are not met.

The way to get around this stereotypical disaster is to have dependencies set in stone;

- *You will __not__ proceed further until problem "X" is resolved.*
- *You may not proceed to level-V commissioning until all level-IV deficiencies have been resolved, quality-checked and signed off by a client representative.*
- *The master contract scope of work says you will provide Tier-IV electric plant operation (including "self-healing" capabilities); you will not get paid until that is demonstrated through thorough testing.*

Construction companies- when faced with either matching the agreed-upon contract or not getting paid- generally will step up and "deliver the goods" rather than not get paid. What is required in the project management triangle is the willingness to forgo a rushed schedule for quality results. Data centers never function well when construction is rushed, no matter what your favorite project manager tells you. And while you may lose some time to make sure the facility is operationally perfect, it is *far* better than having a facility that bankrupts your firm because you cut corners on the schedule, to meet a promise made by executive at a cocktail party.

13 FINANCIAL CONSIDERATIONS OF DATA CENTERS

So you've decided to build a data center. Great!

If you're familiar with finance, you know there are two distinct elements that must be addressed before you go to your financiers, to get funding: capital expenditures (CAPEX) and operational expenditures (OPEX). You'll have a boatload of contractors banging down your door to build you the "perfect" data center... for the right price. How do we determine what that price is?

I've seen some data center professionals say "the ball-park cost will be about $1000/ft^2 or more..." The flaw with this approach is that you may have a power "density" of 40 watts/ ft^2 of Tier I power, or 200 watts/ ft^2 of Tier-IV power... the power and cooling systems are as different as a Model T Ford versus a Ferrari. So we need a model we can use, which will account for not only floor space, but also power capacity and redundancy.

That model was developed by the Uptime Institute in 2008, when W. Pitt Turner IV, P.E. and Kenneth G. Brill published "Cost Model: Dollars Per kW Plus Dollars Per Square Foot of Computer Floor." You can download this at http://www.pducables.com/documents/CostModelDollarsperSqFtUptime Institute.pdf.

The premise of the paper was that the computer square footage (of raised floor) was a separate cost from the power and cooling systems to be installed.

From my experiences, I can attest to the accuracy of these thumb-rules, despite the paper being six years old (at the time I write this).

- Computer floor space cost is approximately $300/ ft^2.
- Storage/"empty space" is $190/ ft^2.

Add to this the "kW" component by desired level of functionality:

- Tier I: $11,500/kW of redundant UPS capacity for IT
- Tier II: $12,500/kW of redundant UPS capacity for IT
- Tier III: $23,000/kW of redundant UPS capacity for IT
- Tier IV: $25,000/kW of redundant UPS capacity for IT

As the paper mentions,

"When one examines data center detailed construction cost breakdowns, the mechanical/electrical systems will account for 70% or more of the construction cost (depending upon density and functionality). This is a distinct departure from office buildings where mechanical/electrical systems are normally about 15% of the total cost of

construction. This fundamental, functional difference between office buildings and data centers is the reason that the traditional office building construction cost benchmarking system of dollars per square foot or per square meter ($/ft² or $/m2) is inappropriate for data center construction."

As an example, a data center feasibility study I performed, necessitated an initial 25,000 ft² data center with room to build another 25,000 ft², with Tier-IV electrical systems and Tier-III mechanical systems, minimal support rooms, with 2,000 kW initially, but modular to allow a maximum power capacity of 10,000 kW (10 megawatts or mW). 10,000 ft² of engineering/support spaces (minimal offices:

Cost of IT floor:	25,000 ft² x $300 ft² =	$7.5M
Tier IV power	$25,000/kW x 2,000kW=	$50M
Extra/Engineering space	10,000 ft² x $190/ ft² =	$1.9M
Total Cost of a small data center:		$59.4M

Notice this is a "strictly business" approach, avoiding any unnecessary costs like office space, lobbies, etc.

Operational Expense (OPEX) Budgeting

Recently, one of the companies I was working with overshot their projected OPEX budget; a series of engineering operations (resulting in significant overtime for the staff) plus some unplanned corrective maintenance, put the company in the red in October, with next to nothing for funds for the remaining two months of the year. A memo came down from on-high, that Senior-VP approval would be required for any expenditure over... $25. We were stunned; that's less than a box of toilet paper for the building bathrooms!

The logical question becomes, "what went wrong with the OPEX budget?"

When it comes to mission-critical facilities, managers typically don't understand realistic operational expense planning for data centers, and underestimate the actual costs. They rely on tried-and-true real-estate management models- for office buildings. Building management organizations such as IFMA and BOMA use a standard OPEX model based on square footage. For example they may assign $1.00 per square foot per year for a premium office, and as little as 20¢ for a typical warehouse. I've seen mission-critical facilities have square-foot maintenance rates (actual) of anywhere from $12 to 18 per square foot, and the Director of Real Estate operations livid that the costs were so out of line for what he expected...

However, mission critical facilities are stuffed with very expensive infrastructure, multiple and redundant generators, uninterruptible power supplies, chiller plants, enhanced fire detection and suppression systems, power distribution units, and are raised floor environment.

And that of course, before you put in your first piece of IT equipment.

The maintenance of all of that equipment is not accounted for by typical OPEX models, and the maintenance costs incurred increase as a function of the amount of IT equipment that is operating in the facility.

So while you may not add any new equipment into your facility in (for example) a two year window, you cannot assume that you're maintenance costs will remain the same in year three, if your IT department is loading progressively more and more equipment into the facility. As the electrical loads increase, the equipment will work harder, requiring more maintenance. This *must* be taken into consideration or you will quickly find your facilities department and running out of money to perform work that you thought was already budgeted for- just like we did.

The Federal Emergency Management Agency (FEMA) had a published white-paper some years back (which I can't locate now), which presented a very good budgeting model and guidelines. The OPEX budget should be ~5-10% of the electricity **cost** used each year.

For example, if you have a data center in the Los Angeles, CA, area that consumes about 1 megawatt (mW) of power, that is 1000 kilowatts per hour, with an average rate (for industrial use) of 17¢/kW/hour. If we take the 10% mark, and multiply the rate by 24 hours, 365 days per year, that cost is $148,920. This becomes are budget for mission-critical operational maintenance costs (parts). As the power drawn by the facility increases, the maintenance budget would also increase, to accommodate the increased maintenance of power distribution, *as well as the cooling and ventilation systems which now must work harder to remove the waste-heat.*

This is the only model I know of, that effectively factors in not only power consumption but also the mechanical system usage to remove the waste heat. While somewhat simplistic, it has proven to be an effective OPEX model

The costs associated with building a data center, can be daunting. Financial questions come up, such as the total cost of ownership (TCO), initial capital outlay and availability. Of course, lower initial capital outlays equal better return on investment, while increased availability enables more consistent revenue streams to the company. Data center management considerations, generally, fall into the question of initial capital outlay, and overall availability. However, as business needs increase in their rate of change, a new element becomes critical to the data center cost equation: agility.

Agility is defined as the ability to accommodate changes in direction of the business- the higher the rate of change, the higher the level of agility necessary, to accommodate that change. Clearly, we can see that the more agile the infrastructure is, the less cost and disruption will be incurred to accommodate change.

While it may sound obvious that we want agility, *how* does one achieve it, when the data center has to be pre-engineered for a specific threshold of power, cooling and space?

This is where the financial calculations come into play.

Every project that is performed, you must know whether it is going to cost the company money or if it's going to be a benefit in financial terms. If you don't have the answer to that question, then you're not ready to start the project. If you're the decision maker and your subordinates to not have these answers, you have a problem. And I don't know about you, but I've always due to business as being in the business to make a profit, not lose money by undertaken the project blindly. Go to sleep

Let's take a real-world example, and see where it takes us. We'll use an apples-to-apples comparison, by running the same assumptions against all three approaches, and only include variables that are specific to each approach (depreciation rates, alterations to cash flow-rates).

The basic assumptions are as follows: the data center that will last 20 years from the day we occupy it. Historical growth is 17.3% per year, in terms of power consumption and floor-space (resulting in a doubling, every 4.5 years). We will also assume that the tax rate to the company is 25%, and straight-line depreciation is over 30 years. Revenues (once the data center is on-line) will start at $10M per year, growing … Initially, we'll only need 2000 square feet of floor space, but by year 5, we'll be somewhere around 4,400 square feet, and by year twenty, we'll be at ~49,000 square feet. Therefore, we'll need a total of 50,000 square feet of IT space.

Option 1: The "C-suite" executives want a flashy facility that they can use as a public-relations tool, to demonstrate "green" technologies, and show off to other IT organizations. This will also include 16,000 square feet of office spaces, a "steel and glass" exterior design, and top-of-the-line décor inside. Estimated cost: $120M. Estimated time to completion is three years. This will be referred to as the "Taj Mahal" data center, to reflect the ostentatious nature of the facility.

Option 2 is a "lean and mean" data center of 50,000 square feet, modular and scalable infrastructure. It will be essentially unmanned, with no specific offices and minimal décor. It is designed to be unremarkable in appearance, so as to be inconspicuous to interlopers who may wish to harm the corporation. Estimated time to completion is 1 year, at a cost of $40M.

Option 3 is to simply lease a facility from a company that specializes in collocation services; with a lease, we bypass the time and upfront costs of constructing a new data center, at the price of increasing real-estate costs, dictated by the business and real-estate markets.

While we will later dive into the details of how such calculations are accomplished, the financial results are in the following graph. The numbers on the left axis represent net present value (NPV) of each approach in millions of dollars; the horizontal axis represents the years.

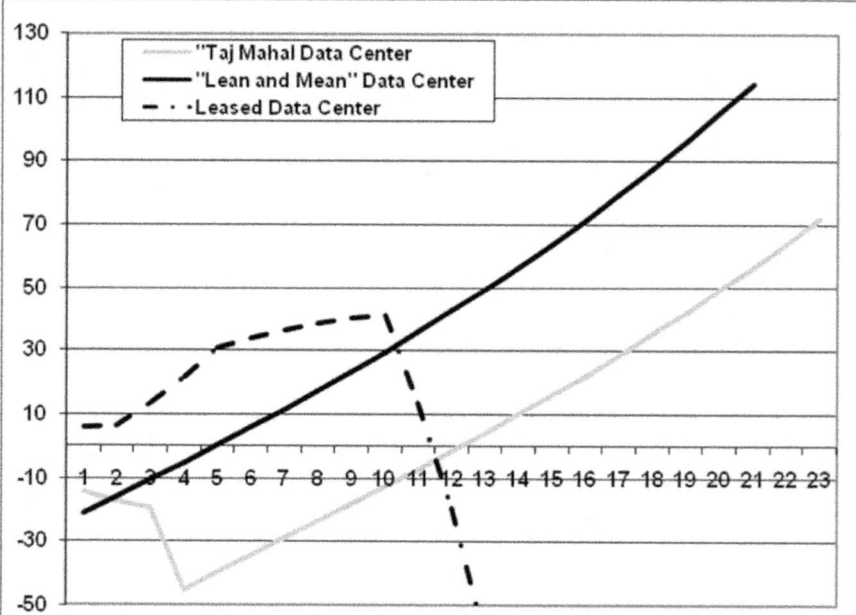

As you can see, the "Taj Mahal" approach is very expensive, and is mortally flawed by the high initial capital outlay and time required to complete. It doesn't reach the payback period for almost twelve years, and has a cumulative NPV of only $140M.

The "Lean and Mean" data center, due to a construction cost only 33% of the Taj Mahal, as well as a relatively fast construction time, is able to capture incoming revenue streams much earlier, and reaches its payback period in only four years. Cumulative net present of the lean-and-mean data center is ~$841M over its twenty year life.

The leased facility looks very good at first, due to the lack of capital outlay required, and effectively no delays due to construction. It's a good approach to take, if you lack the $40M (much less $120M) required to build your own facility. The leased facility, however, has one main factor against it: monetary inflation. Between this, and the projected growth of power and footage, at year five, when we have to double the amount of real-estate available, our return on investment starts to diminish, and at year ten, when we have to double the footprint again, return on investment is negative. Total net present value of this approach, by only year 13, is -$60M, and it only gets worse from there!

There is something to seriously consider on the leased facility approach. Realistically, a data center has a useful life of approximately 10 to 15 years. When we say useful life, we mean the amount of time before major it infrastructure upgrades must be performed to allow the facility to continue to support the ever-increasing demands of IT equipment with regards to both power, cooling and rack power-density. Remember, load growth almost always exceeds the expectations of the design engineers and when did you have the constantly changing expectations of IT departments, forcing used to cram more power and cooling into a smaller area, by year 15 you should reasonably expect that your infrastructure will not have sufficient flexibility continue to accommodate your enterprise IT-department. So when you build a new data center, you have the reasonable expectation that the facility will function without major issues for 10 to 15 years, right?

As I mentioned earlier, agility is an extremely seriously consideration; lack of it costs BIG! And after 10-15 years, your facility will lack the agility needed to accommodate your needs. And at the end of that time, when you go to perform major engineering upgrades, you will incur significant capital costs as well as serious risks to your IT operations.

How do you avoid this?

This is where the leased facility really shines-it allows you to move into a new facility where you can fully utilize the electrical and cooling support systems until their useful end of life, and by year 15 move into a new leased facility or even a new data center of your own construction while avoiding the costs of engineering upgrades in the meantime. That is to say, you can use the leased facility for its useful life, and then move on to renew leased facility while avoiding the intermediate costs and risks associated with the engineering upgrades necessary for continuous operations. In this way, you maximize your business agility and avoid the capital funding issues needed to force an owned facility to keep up.

It is also very useful for the company which is short of a capital project funding, yet can justify and generate sufficient cash flow for leasing.

Initially, looking at the NPV charts, it does not appear to be an ideal situation, but when you dig deeper into the overall picture, this turns out to be a very attractive option.

I want to bring up an interesting point, which the NPV graph forces us to consider: up until now, an enterprise data center is generally considered a cost center. Yet, when considering the graph and NPVs of each approach, it is obvious that a data center can actually be a *profit* center. How we determine this is in the same manner that I used for the NPV calculations above.

In other words, we have to look at the expected revenues that the new facility will generate, less the fixed and variable costs associated with that facility, and you now have the ability for an honest assessment of your facility, with regards to your business model. The difficulty, of course is not in finding out what your fixed and variable costs are that are associated with the data center. The costs are the *easy* part- as we have already shown.

*The hard part is figuring out what monetary value your **data center *provides***.*

To get down to the basics of what the value of your data center provides, you must have a logical and realistic model which will tell you what each application brings to the company. It's not enough to say "this will enhance the customer experience," because that doesn't mean anything. You have to assign a monetary value to each enhancement, to each application, to each offering that is implemented. Obviously, we have to start with the basics, and track the changes in revenues that each application generates. In the case of the software suite for example, we would look it either a total enhancement of revenues with a total savings and costs associated with the implementation of that software.

Without having a way to measure the monetary value for each enhancement or application or offering or whatever, you simply have no tools to find out what brings money to the company and what takes money *from* the company. If you're a small mom-and-pop shop, you have to take chances and to guess at what will work- flying by the seat of their pants in a primitive airplane. But if you are an enterprise IT organization, spending tens of millions of dollars a year on new applications and goodies for your customers, you'd better have a very good idea of what they were going to like and what they're willing to pay, for those goodies. Flying by the seat of your pants in a jumbo jet is not acceptable.

Not only will having this tool allow you to make more informed decisions, it will also bring about a level of accountability that is lacking without the tool. In other words those people that waste money will be found, and the situation can be corrected. Those people that come up with money making ideas can *also* be found, to the betterment of the company.

*Oh, I **know** that this particular observation is not going to make me the most popular person amongst IT professionals, but it is something that I have noticed over the years.*

What I propose is no small feat; in fact, most IT people would tell you it's impossible. But without the proper tools to assess the value of a project, it becomes impossible to make consistently sound management decisions!

To be clear, it's *not* easy to make a data center a revenue center, but it is certainly achievable, with proper management, planning and view each project with the question that MUST stay at the front of the decision-makers' mind… "Does it make financial sense?"

Let's revisit that graph, and see how we arrived at our conclusions.

Going back to the NPV charts above, this chart was for a real company using actual estimates assigned to a planned facility. The company already owned the land; the costs were strictly assigned to construction. Engineering architecture was to be built for 100% of expected power and cooling capacity from day-1. This means that, in the early years, the unused power and cooling capabilities were wasted capital dollars, commonly referred to as "stranded capacity." The costs of such capacity were already incurred, but could not be used in any meaningful way.

INFRASTRUCTURE APPROACHES: LEGACY THINKING vs MODULAR AND SCALABLE

The "lean and mean" data center assumed the same plot of land was used, but instead relied upon local construction costs of a raised-floor environment,

A key feature of the lean and mean data center, is that it takes advantage of tax benefits, which allow accelerated depreciation of assets. The engineering architecture allows us to start small, and bring in additional generators, chillers or UPS systems, as business needs require. This bypasses the costs of stranded capacity, and maximizes business agility.

Another important feature of this approach, is to use pre-engineered equipment designs for our data center. In other words, we use systems that allows is to start with a single modular component, and allows us to add more modules later, in a like fashion. Such pre-engineered solutions offer a level of uniformity and scalability that the legacy approach (of the "Taj Mahal") lacks. These systems, since they're designed, built and tested at the factory, can be installed in the facility like an erector set. As such, they're not considered "building improvements," which raise the taxes on the property and must be depreciated over a 30-year lifespan. Instead, they're considered "computer equipment," with a much shorter life, and greatly accelerated depreciation schedule.

In the following example, we assume that the business revenues will be $10M for year 1, and grow at 15% annually, over the next 20 years. We'll use $10M of infrastructure in both the modular and legacy approaches, but the modular will start out with only $2.5M of cost, and an additional $2.5M will be included each five years thereafter. Depreciation will be straight-line, over a 7-year span (like computer equipment is). The legacy system, since it's a building improvement, will feature $10M of cost up-front, and be depreciated straight-line over a 30-year span. [Please note, however, that while building improvements are 30 years, the typical life of the equipment is only 20 years, resulting in tax write-downs at the end of 20 years, resulting in further financial loss.] Inflation is assumed to be 5%.

As you can see in the following graph, the difference in NPV is $4.36M- nearly half the cost of the infrastructure itself! With the modular approach, the stranded-capacity issue is avoided, and we're able to capture almost all of the depreciation. A significant issue that we avoid in the modular system, is having to write down 10 years' worth of depreciation on the balance sheet.

Remember, this is a very simple example, so as they say, "Your mileage may vary." If, for example, you discover that your IT department grows at a slower rate than forecasts predicted, the costs of additional infrastructure can be deferred to a later date, again avoiding stranded capacity and capital expenditures, until necessary.

The purpose in including this graph, of course, is to illustrate that things are not *always* as they seem; modular, scalable systems offer some cost savings in the overall equation. What cannot be quantified in the NPV calculations, however, is the ability to scale your infrastructure on an as-needed basis. If your forecasts are wrong and you need to scale up sooner, than there is little cost associated with doing this in advance of what you anticipated. On the other hand, if growth does *not* occur as you expected (for whatever reason) you are now paying for unused capacity with the legacy approach; stranded capacity.

Infrastructure NPV- Modular vs. Legacy as a Function of Costs, Depreciation and Net Revenues									
Year	Business Revenue	Modular Cost	Mod. Deprec.	Legacy Cost	Legacy Deprec.	Net Rev. (Mod.)	NPV Modular	Net Rev. (Legacy)	NPV Legacy
1	10.00	-2.50	0.36	-10.00	0.33	7.86	7.86	0.33	0.33
2	11.50		0.36		0.33	11.86	11.29	11.83	11.27
3	13.23		0.36		0.33	13.58	12.32	13.56	12.30
4	15.21		0.36		0.33	15.57	13.45	15.54	13.43
5	17.49		0.36		0.33	17.85	14.68	17.82	14.66
6	20.11	-2.50	0.36		0.33	17.97	14.08	20.45	16.02
7	23.13		0.71		0.33	23.84	17.79	23.46	17.51
8	26.60		0.71		0.33	27.31	19.41	26.93	19.14
9	30.59		0.36		0.33	30.95	20.95	30.92	20.93
10	35.18		0.36		0.33	35.54	22.91	35.51	22.89
11	40.46	-2.50	0.36		0.33	38.31	23.52	40.79	25.04
12	46.52		0.71		0.33	47.24	27.62	46.86	27.40
13	53.50		0.71		0.33	54.22	30.19	53.84	29.98
14	61.53		0.36		0.33	61.89	32.82	61.86	32.81
15	70.76		0.36		0.33	71.11	35.92	71.09	35.91
16	81.37	-2.50	0.36		0.33	79.23	38.11	81.70	39.30
17	93.58		0.71		0.33	94.29	43.20	93.91	43.02
18	107.61		0.71		0.33	108.33	47.26	107.95	47.10
19	123.75		0.36		0.33	124.11	51.57	124.09	51.56
20	142.32		0.36		0.33	142.67	56.46	142.65	56.45
Total			9.29		6.67	1023.72	**541.40**	1021.10	**537.04**

Going back to our real life example, IT management made a final decision as to which approach to take; since they failed to perform any financial calculations and lacked sufficient concern for obtaining sufficient capital funding, they decided to go for the Taj Mahal.

Even before they broke ground on the facility, they experienced difficulties getting sufficient capital funding. Because the C-suite executives were so fixated on their fancy offices and flashy IT spaces, when the need to reduce costs arose they started undercutting the key infrastructure pieces that would keep their IT equipment running.

Predictably, their efforts to reduce costs by 25% in turn led to increased costs in other areas, causing a variety of complications in the engineering design, leading to serious cost overruns. Ultimately they paid 38% *higher* than the initial bid, due to a failure of leadership, technical understanding or oversight, and not staying focused on what really matters in a data center. The actual cost was double what they projected with all of their cost "savings."

The weakness with this entire approach was that IT executives failed to stay focused on the systems that would support their business needs and instead became concerned with peripheral issues that did not matter to the business model. The lost focus on the things that were really important to the company, and the company paid the price.

The proper approach would have been to go with the lean-and-mean facility, because this particular company has a history of staying in their facilities for extended period of time, 20 to 30 years. Going back to the NPV charts, it becomes clear why this approach is best.

FINANCIAL CONSIDERATIONS- OTHER

Cost Center vs. Profit Center

Historically, data centers have always been regarded as cost centers. This clearly comes as a delineation between IT operations and facilities management. That is, IT departments typically regard their costs to include the purchase or lease of the server, installation and configuration costs, and management costs of the server itself, *minus the costs of the underlying infrastructure systems*. Put another way, the typical mode of business is that the power and cooling are "free" or merely someone else's budgetary problem to deal with.

While Moore's Law predicts a doubling of computing power every 18-24 months, there is a corresponding growth in power consumption by IT equipment over the same period, albeit at a lower rate of compounding. In practical terms, this translates into IT power and cooling demands escalating at a rate that often quickly exceeds the speed with which additional power and cooling capabilities can be augmented, to handle the increased demands. Server compute performance has been increasing by a factor of three every two years. This means computational performance increased by a factor of 27 between 2000 and 2006. Energy efficiency has gone up as well, but the at-the-plug power consumption has still risen by a factor of 3.4 (Brill, 2007). That works out to 23% load-growth year over year!

The reality is that the costs for additional power distribution and cooling *must* be accounted for in the IT Total Cost of Ownership model. In other words, the power and cooling support systems required are a fixed cost associated with the power usage of each server installed in the data center; if those costs are allocated properly to the critical facilities OPEX and CAPEX budgets, a huge bottleneck to the operational effectiveness of enterprise data centers will be eliminated (we'll get deeper into this when we discuss ITIL).

If the critical facilities department is able to operate under the average infrastructure costs associated with the needed servers, they will then be operating under budget, and over if the inverse is true.

This leads to a prickly subject; who should critical facilities staff report to? Due to the inherent link between critical facilities budgeting and IT departments, the critical facilities staff and budgets must work as a subsidiary to IT executive management, versus the (typical) corporate facilities team. The budgeting issue is one aspect, but another is that the critical facilities staff must closely work with their IT counterparts, in a collaborative environment. With such communication and with earned trust, the resulting team is far more effective in resolving issues, planning for the future and responding to customer needs, than by operating in silos.

14 INFORMATION TECHNOLOGY INFRASTRUCTURE LIBRARY (ITIL)

ITIL is the gold-standard of IT business operations models, developed in the United Kingdom back in the 1980's to standardize how IT operations should be managed. It copies the Charles Deming continuous improvement cycle, and is essentially a communications model that allows different IT operating groups to effectively work with each other, through vehicles like Continual Service Improvement, Service Operations, Service Design, Service Strategy and Service Transition. I must stress that these programs are essentially *communication models,* proposed as recommendations for enterprise organizations to follow.

ITIL is not without its flaws; it lacks industry-standardized key performance indicators, it's a "one size fits all" approach (requiring adaptation to each company that utilizes it, and it while "everyone knows" ITIL improves service delivery, the mentioned lack of KPIs relegates this belief to a leap of faith, as no data is available to back up this belief.

ITIL, it is believed, reduces operational costs and technical problems, by standardizing IT operations, but again there is no data to back this up.

Anecdotal evidence strongly suggested that IT departments were largely uninterested in including Critical Facilities staff in the strategic planning portions of ITIL, resulting in insufficient power and cooling availability, leading to decreased reliability, higher operational costs for emergency "catch-up" system upgrades, and decreased profits.

I was therefore compelled to research this topic, to explore whether ITIL offered the returns that are generally accepted as a given. But as someone focused on infrastructure- power, cooling, fire & life-safety, I approached it from the question of overall availability and reliability of the affected data centers. The research questions were quite simple:

    1. Does ITIL have any effect on reliability of the mission-critical

facility?

2. If you use ITIL in your enterprise IT department, are the Critical Facilities staff included in the strategic planning?

3. How many outages have you had in the last five years at your primary facility, due to cooling or power issues?

Quantifying the efficacy of a communication model is nearly impossible; it's like trying to measure "the telephone game," that children play. How do you measure the degradation of the message as it passes from one person to the next?

We had the independent variables: use of ITIL (yes or no), and does your Critical Facilities (CF) staff get included in the ITIL planning process? The dependent variable was the # of outages. By analyzing the results, the following chart results:

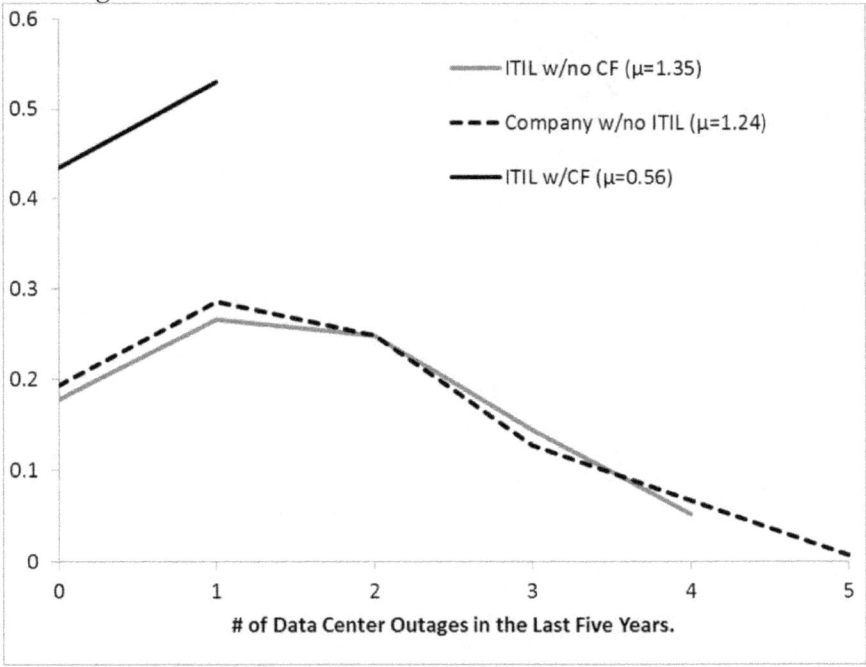

Those companies that used ITIL without including their CF people in the strategic planning of IT operations (most companies) actually saw their overall reliability slightly *decrease* with the implementation of ITIL in their organization. But those companies that *do* include the CF staff in strategic planning have reliability 275% higher than those that do not.

What this means- in realistic terms- is that CF staff must be included in the strategic planning model of enterprise data centers, for those data centers to have a realistic chance of performing as intended.

Indeed, analyzing the data further, IT departments strive for "five nines" of performance; 99.999% uptime. The results clearly indicate that this is impossible without the CF staff being directly involved in the strategic planning and management of the mission-critical facilities.

For a guy who started out as a technician fixing machinery, when someone would say "it's all about communicating," I would shake my head in disbelief. Now, some 30 years later, the data clearly show that this IS the truth. When we plan together and work together as a team, is when the *real* magic happens! The results of this research has literally changed how I view my work, and how I view data centers and managing engineering teams.

Hopefully, it will help you in your efforts!

The full dissertation on this research is in Appendix II, at the back of the book.

APPENDIX I: TIGER-TEAM PROPOSAL

This proposal was created in response to a need to have skilled technicians service a portfolio of mission-critical facilities over a large geographic area. This is provided as an example of a flexible staffing model when 'normal' proposals are inadequate. This also demonstrates how budget calculations can be used in unusual circumstances.

Critical Facilities "Tiger-Team" Proposal

Summary:

1. This addresses the need to create a critical facilities technical team, to provide technical oversight of preventative and/or corrective maintenance of all critical facilities portfolio infrastructure systems.
2. Provides team of in-house experts to deal with unplanned events.
3. Removes "tribal knowledge" currently held by outside contractors.
4. Builds sense of ownership and responsibility of in-house technicians.

Background:

The current critical facilities management model is essentially the model that is typically applied to the management of typical real-estate portfolios.

*For a typical real-estate portfolio, the traditional approach is to create a budget, which relies on contractors to perform mundane/non-technical tasks, such as gardening/landscaping functions, janitorial services, security, simple electrical troubleshooting, basic plumbing repairs, etc. Due to the lack of complexity of such facilities, specialists are rarely needed, unless there is a catastrophic failure of a system (such as a ruptured sanitary line). Some maintenance functions are performed in an on-going basis, for things like HVAC filter change-outs, minor customer-service calls like plugged toilets or broken latches, and so on.

In this situation, the technical knowledge necessary for such maintenance functions are relatively simple, and maintenance technicians can be hired and trained rather easily. Should such a technician make a mistake, the costs are minimal, consisting largely of inconvenience and the occasional replacement of damaged components, with essentially no impact to business continuity. As a result, the technicians and other support staff have little bearing on the survival of the company.

A mission-critical facility is, by definition, a building that shelters and supports computers and other ancillary equipment, necessary for continued business operations. Therefore, in a data center, human error has a direct impact on the survival of the business. According to the Uptime Institute, over 60% of downtime is due to human error.

Keeping the above in mind, a core group of highly-trained, well-managed critical facilities technicians are an absolute necessity, to maintain the high-availability data center. Indeed, per the Uptime Institute white-paper, "Critical Environment Staffing Considerations" (Rick Schuknecht, Data Center Operations and Management Specialist):

- The quantity and quality of data center staffing are both key to achieving continuous availability.
- Corporations that are serious about uninterrupted uptime will treat staffing as a cornerstone of their plan to achieve it.
- Achieving excellence in staffing at key critical facilities will require hiring qualified personnel at rates higher than their counterparts with the same job title who are assigned to non-critical facilities.

Unfortunately, the staffing of the critical facilities is essentially non-existent. Core services are provided by external vendors who may lack the allegiance and ownership that such a role demands. Those vendors are managed by Facility Managers that are familiar with typical real-estate management portfolios, but unfamiliar with the increased technical understanding required, and thus are unable to exercise a sufficient level of supervision, to make sure the critical nature of the locations are sufficiently addressed.

Those external vendors utilized by generally lack the specialized skill-set required in a critical facilities environment, instead having more generalized (occupied) building trade skills or construction skills (please note that this a very common problem). With insufficient skill-sets, they tend to miss key issues, such as hot-aisle/cold-aisle containment of IT floors (crucial to maintain a hospitable computer environment), and strategic planning and foresight for future issues that will require infrastructure upgrades.

Even worse, a typical problem is that vendors will not be sent out for periodic training on the newest critical facilities technologies and equipment they may see, or the methodologies developed to successfully oversee the facility under their control. The problem arises from the debate of whether the company pays the training and transportation/lodging costs, or the vendor, as part of the contract.

The net result (as expected), is a lack of control within the critical facilities physical environment (data center floors), with the follow-on result of hot-aisle/cold-aisle containment, and a lack of a coherent overall strategy for production floor IT equipment installations. These management weaknesses, in turn, result in lower than expected cooling efficiency and increased operational costs, while leaving critical IT equipment exposed to overheating in certain areas.

Conclusion from Current Situation:

Overall, the current critical facilities management model is unsatisfactory. It is my recommendation that a core team of facilities technicians be hired and trained (on-site, using the already-developed training program) on critical facilities architecture, philosophies of operation, specific critical facilities equipment- UPS systems, batteries, HVAC in the data center, emergency procedures, normal electric plant operations, safety system, building monitoring systems, etc. After these technicians are hired and trained, they can be utilized to man the primary critical facility locations of AAA and BBB data centers, as well as supplement the manpower needs at other facilities, in the event of a casualty or during major work.

Due to the increasing importance of both the AAA and BBB data centers as co-location facilities, it is my recommendation that a 24x7 coverage of both facilities be maintained, to monitor those facilities for unusual conditions, with the Tiger-Team being on-call in the event of emergencies. By using this approach, vigilance can be maintained at the data centers, with the ability to quickly respond to unusual events, not only at the data centers, but at any location within the critical facilities portfolio. This step will allow the company to accomplish three goals:

1. Reduce risk of failure in critical facilities, by having in-house experts on our systems, who can monitor and track system issues across all locations,

2. It will offer a level of resilience and flexibility in scheduling, especially during unplanned events, where a "tiger team" can assist the local manager with repairs/recovery.

3. Reduce the manpower costs associated with such staffing by having direct employees, rather than hiring external union workers, and pay commissions to the staffing company that acquired them.

Proposed Critical Facilities Strategy:

The success of the Tiger-Team depends on a multi-faceted approach, to integrate technological best-practices, with comprehensively trained technicians.

1. Implement best-practices critical facilities architecture, with established goal of Tier-IV electrical design (allows concurrent maintenance of all systems, and fault-tolerant) with Tier-III cooling systems (allows concurrent maintenance, with N+1 redundancy, minimum).
2. Rely on proven PLC automation technologies, to allow critical-facilities architecture to be self-adapting in unusual/emergency events. This reduces risk of human error (causes 70% of all critical-facility outages, per Uptime Institute), and reduces physical risk to personnel in case of an emergency.
3. Take advantage of IT capabilities, to have remote monitoring/tracking of critical facilities conditions and efficiencies.
4. Leverage those IT capabilities to have an on-call "duty engineer" from the Tiger-Team, who can be emailed via Blackberry, any unusual situation alerts, and can remotely inspect the system. The duty engineer can then determine whether an event is minor, or if it is an all-hands event, and take appropriate measures
5. Utilize "extra eyes" wherever possible, to provide input to any unusual conditions which may be developing (but would not be required to physically touch any equipment).

Plan of Execution:

6.   Initial hiring of the tiger team should consist of the following:
1. Two junior electrical technicians
2. One junior mechanical technician,
3. Two senior technicians
   1. One electrical specialist,
   2. One mechanical specialist

7.   Training should consist of a combination of
1. Studying critical facilities whitepapers, research and other publicly-available information on best-practices within the critical facilities environment,
2. External training by vendors for equipment used within the critical facilities portfolio, including:
   1. Liebert UPS systems
   2. Powerware UPS systems
   3. Caterpillar Generators,
   4. Chillers
      1. Theory including refrigeration cycle
      2. Practical applications
      3. Troubleshooting
   5. Switchgear- ASCO, Eaton, G.E., etc.
   6. Fire Suppression System (Halon, FM-200, etc.)
3. Free External training provided by APC (and other IT-based firms) that offer sound high-level understanding of critical facilities and their system interrelationships.
4. Training on regulatory issues, including
   1. SCAQMD
   2. CAL-OSHA
   3. FED-OSHA
   4. Applicable National Fire Protection Agency Codes, for
      1. Electrical Safety
      2. Fire system requirements in IT environments
      3. Electrical Engineering Standards
      4. Switchgear Testing
      5. Fuel-oil systems and safety
5. "Practical Applications," including discussions and demonstrations of operational knowledge and casualty-response of the following:
   1. Electrical Systems

       2.   Mechanical Systems
       3.   Fire and Life-Safety Systems
       4.   Power installations
       5.   Typical Maintenance Operations

8. Utilize a training plan that allows for frequent reviews, examinations and opportunities for learning, and measures the progress of the tiger-team technicians/specialist against those benchmarks established in the training program.

    1.   Time to complete training: 6+ months
    2.   If candidate is unable to complete, or demonstrates that they cannot handle the task, they may be reassigned to another department

9. Implement automated paging system, utilizing Blackberry's, to alert Tiger-teams of any unusual/emergency conditions at the monitored data centers.

    1.   Paging system should include alerts of out-of-spec conditions in the following systems:

         1.   Power distribution
         2.   Generator status
         3.   UPS status
         4.   Cooling system status
         5.   Fire/Life-safety system activations/alarms

    2.   Set up Tiger-team members with ability to perform remote-monitoring of all data centers via company-issued laptop computers, with appropriate IT-designated security protocols

Implementation:

1) After tiger-team is fully trained and in-service,
2) Scale back on-site engineering staff to two person per location, day-shift only
   a) On-site staff will be present to
      i) Work with project/engineering personnel, as needed.
      ii) Perform power installations/removals for IT equipment
      iii) Perform minor maintenance as required by Facility Manager
      iv) Be first-responder to unusual conditions/alarms, and take corrective actions to put plant in a stable condition, until Tiger-team personnel arrive.
3) After-hours monitoring will be provided by security guards who perform rounds
   a) Guards will provide visual observation only
      i) Guards will call on-call Tiger-team member for any unusual

conditions/alarms detected

    ii) Guards will be trained on basic electrical safety, so they know to touch nothing

    iii) Guards will not open any doors, lift floor-tiles or any other manual functions, which might risk their safety.

b) On-call Tiger-team member will be provided with "duty cell phone", and be able to remotely monitor conditions of any facility.

c) On-call Tiger-team member will be responsible for initial engineering response to any unusual conditions, and be expected to perform initial analysis/triage of problems within any facility

    i) Must be able to take immediate corrective action for all reasonably conceivable events that may risk the facility.

    ii) Must be able to take immediate actions to put the plant in a stable condition (until help arrives, if needed) in all reasonably conceivable alarm/emergency scenarios.

Required Tools/Equipment:

Initial equipment will consist of:

10. Basic electrical and mechanical tools
11. Mandated Personal Protective Equipment (PPE)
12. OSHA-approved Lock-out/Tag-Out kits
13. Training Materials (as required)
14. Specialized tools as deemed necessary to perform specific functions, to be procured on an as-needed basis.

Cost Analysis:

1) Estimated current costs for existing 24x7 coverage:

a) BBB Operations Center

    i) Critical Facilities Contractor coverage- $840,000

      (1) This includes only six engineers,

      (2) BBB is understaffed to maintain required coverage.

b) AAA contractor coverage- $1,775,172

    i) Day-shift Electrical Contractor- $280,000 (estimated, billed to IT Department)

    ii) Off-shift Electrical Contractor- $611,520 (enhanced coverage replacing CRITICAL FACILITIES CONTRACTOR)

    iii) Day-shift Mechanical Contractor- $114,612

    iv) Off-shift Mechanical Contractor- $769,040 (enhanced coverage replacing)

c) CCC Costs (estimated, when construction is complete) -$840,000

    i) Based on similar size to BBB facility,

      (1) It will have more sophisticated monitoring system,

(a) Increased mean-time between failures,
(b) Decreased mean-time to recovery
(2) Have more specific system knowledge requirements to properly maintain
(a) Due to more sophisticated systems
(b) Cutting-edge technology

2) Estimated labor costs for Tiger-Team- $930,000/year
   a) Two Senior Technicians (Base Pay ~$35-45/hour)- $480,000 loaded rate
   b) Three Junior Technicians (Base Pay ~$20-30/hour)- $450,000 loaded rate

3) Overlap of Teams:
   a) Tiger-Team recruitment and training time = 1 year
     i) $930,000
   b) AAA Staffing during this time = $1,775,172
   c) Irvine Operations = $840,000

4) After ~ 1 year, begin phase-out of Precision Electric, ACCO services
   a) Rate determined by how quickly Tiger-Team is to take over tasks
   b) For simple cost analysis (in graph, last page), assumption is cost ↓ 33%/year.
   c) Costs assume 5% Inflation (Pay Raises) per year.
   d) As graph shows, when comparing Tiger-Team implementation costs versus current growth projections (AAA only), total costs decrease at end of year 2, with successively higher savings as time goes on.
     i) BBB Operations Center savings would make this even more dramatic.
     ii) Tiger-Team would cover all critical facilities instead of two primary sites and no coverage at smaller locations.

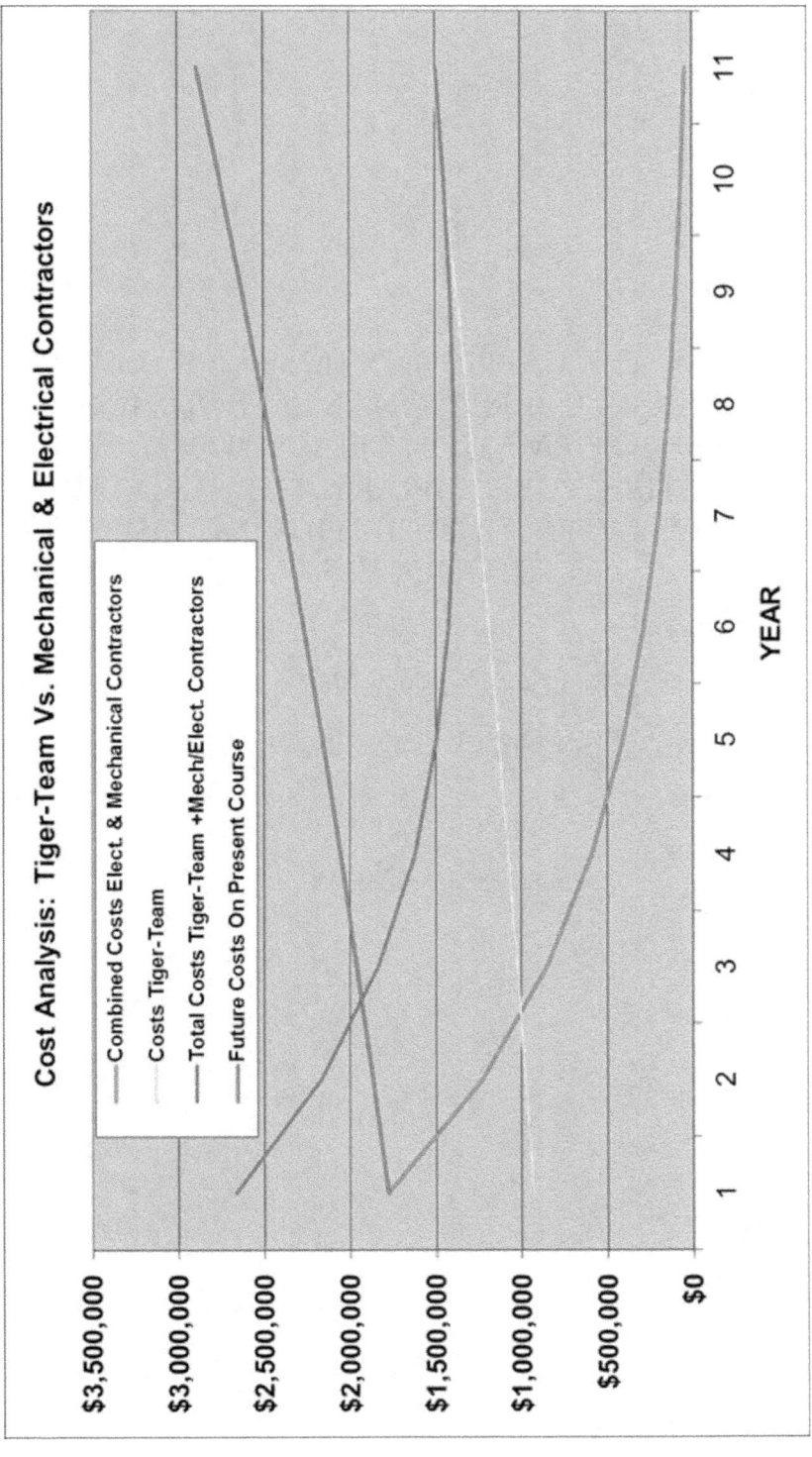

Cost Analysis: Tiger-Team Vs. Mechanical & Electrical Contractors

— Combined Costs Elect. & Mechanical Contractors
— Costs Tiger-Team
— Total Costs Tiger-Team +Mech/Elect. Contractors
— Future Costs On Present Course

YEAR

APPENDIX II:
"MEASURING OPERATIONAL EFFECTIVENESS OF INFORMATION TECHNOLOGY INFRASTRUCTURE LIBRARY (ITIL) AND THE IMPACT OF CRITICAL FACILITIES INCLUSION IN THE PROCESS."

Eric A. Woodell, D.Sc.

Robert Morris University, 2013

Supervisor: David Wood

Abstract

Information Technology (IT) professionals use the Information Technology Infrastructure Library (ITIL) process to better manage their business operations, measure performance, improve reliability and lower costs. This study examined the operational results of those data centers using ITIL against those that do not, and whether the results change when traditional facilities engineers are included in the process. Overall, those IT departments using ITIL processes had no statistically significant improvements when compared to those who do not. Inclusion of Critical Facilities (CF) personnel in the framework offered a statistically significant improvement in their overall reliability of their data centers. Those IT departments who do not include CF personnel in the ITIL framework have a slightly lower level of reliability than those who do not use the ITIL processes at all.

Determining Whether the Inclusion of Critical Facilities Support Personnel in the Information Technology Infrastructure Library (ITIL) Process Yields Meaningful Improvements for Enterprise Information Technology Department

Chapter 1: Introduction

1.1 ITIL Background:

As early as the 1980's, the growth of Information Technology (IT) indicated that there was a need for standardized procedures for the deployment of IT equipment in protected environments like data centers. In response to this need, the Government's Central Computer and Telecommunications Agency in the United Kingdom developed a set of standard processes to fill this gap. These processes, which focus on the IT Service Delivery methods, are now known as the Information Technology Infrastructure Library, or ITIL, which is based on the "Deming Cycle" of continuous improvement. ITIL is literally a set of five books that provides recommendations and best-practices for IT with regard to equipment installation, maintenance, problem-resolution and overall management.

An interesting aspect of ITIL is that while it provides basic management guidelines and charts, the expectation is that the actual implementation is customized to meet the needs of the individual firm, with no specific rules. Thus the very flexibility which makes it attractive for adoption results in inconsistent implementation of the process and meaningful measurement of performance is problematic; the very one-size-fits-all approach does not offer firmly established measurement techniques, only suggestions. Or as one analyst put it, "it is a descriptive framework, not a belief system. ...some assembly required" (Nichols, 2007). Suggested measurements of the ITIL process includes Key Performance Indicators (KPIs) to measure the effect of the ITIL implementation project (Litten, 2008), but the lack of standardization of those KPIs makes industry-wide benchmarking unlikely, since no two companies measure the same things in exactly the same way- a logical result of the generic nature of ITIL. As a result, the idea of 'being ITIL compliant' is made irrelevant (Bruton, 2005).

The ITIL process recognizes that "traditional facilities" like buildings, power, cooling and water are necessary for the successful operation of IT assets, but this is merely lip-service, since these factors are studiously ignored throughout the ITIL suite. In other words, ITIL is an IT matter only, with no consideration of business strategic, commercial and political matters (Bruton, 2005).

1.2 Ramifications for the IT Environment:

The ever-increasing reliance on Information Technology is concisely captured by Moore's Law, which proposed that the number of processors on a computer chip would double every 18-24 months (Moore, 1965). And while the doubling of processors results in a corresponding increase in computational speeds, this increased speed comes at the price of increases in power demanded, and the waste byproduct of power consumed is heat. The practical implication of Moore's Law is that, while the power consumed by that IT equipment does not double over the same period, it continues to increase at a predictable, compounding rate, with serious long-term implications for the facilities in which the IT equipment is housed- the data center.

1.3 Data Centers:

A data center is a centralized repository, either physical or virtual, for the storage, management, and dissemination of data and information organized around a particular body of knowledge or pertaining to a particular business (Godinho, 2000). Enterprise data centers are specially constructed with dedicated cooling systems (referred to as HVAC) and power distribution systems, which incorporate layers of redundancy to guarantee environmental conditions and power are uninterrupted. The data center is designed with sufficient protective systems to compensate for service interruptions, and maintain operational readiness for the Information Technology customers. For the business which relies heavily on IT, large-scale data centers are a logical choice for protecting their core business. A key consideration in data center design is that the high levels of redundancy and reliability that are designed into the facility are planned to support a specific amount of power, which must be forecasted before construction can start.

While Moore's Law predicts a doubling of computing power every 18-24 months, there is a corresponding growth in power consumption by IT equipment over the same period, albeit at a lower rate of compounding. In practical terms, this translates into IT power and cooling demands escalating at a rate that often quickly exceeds the speed with which additional power and cooling capabilities can be augmented, to handle the increased demands. Server compute performance has been increasing by a factor of three every two years. This means computational performance increased by a factor of 27 between 2000 and 2006. Energy efficiency has gone up as well, but the at-the-plug power consumption has still risen by a factor of 3.4 (Brill, 2007).

1.4 Overall Implications:

In short, IT power and cooling demands are literally growing faster than the "traditional facilities" people can increase the capabilities of the data centers- resulting in new risks for the entire IT enterprise. A 2006 survey pin-pointed this as a major issue, where "almost 40% of respondents reported that they had run out of space, power, or cooling capacity without having sufficient notice. This can result in delaying provisioning of new initiatives or paying more for additional resources to host those systems. These survey results are indicative of a burgeoning problem in data centers today where there is no mechanism in place for proactively managing and optimizing resources." (Aperture Research, 2006) A similar survey in 2009 confirmed "[i]n many data centers, power and cooling systems are stretched to the limits, quickly approaching capacity. Power and cooling already account for more than 50 percent of IT spend. Power demands continue to escalate. High-density servers generate more heat than ever—heat that has to be addressed by overstressed cooling systems that consume more power that in turn generate more heat. In this vicious spiral, more than 40 percent of data centers will run out of power capacity..." (Spears, 2009) As of May 2011, this issue is still of primary concern, when the Uptime Institute indicated that 36% of data centers will run out of space, power or cooling in 2011-2012 (Stansberry, 2011).

To upgrade in-service data centers for increased power and cooling capacity often requires years to plan and execute, and significant expenditures of capital In economic terms, the resulting increase in server power consumption now burdens data center operators with increasing electricity and site infrastructure costs for every server they deploy (Brill, 2007). Failure to sufficiently keep up with the power and cooling demands of increasing IT loads results in decreased reliability and redundancy of the data center, with outages being the result.

This is far more frequent than one would think. "Over the next five years, power failures and limits on power availability will halt data center operations at more than 90% of all companies. It's a fact of life: Outages happen. If you've dodged the bullet thus far, you're part of a fortunate minority. A recent AFCOM survey found that 81 percent of respondent had experienced a failure in the past five years, and 20 percent had been hit with at least five failures. Of those outages, more than 80 percent were due to either external or internal power failures." (Miller, 2006) The overall situation worsening and the consequences for failure to resolve these issues are severe. "The data centers shared environmental and human support services are complex. Without data-center-management coordination, efficiencies achieved through sharing are quickly replaced with extended downtime and mean time to restore." (Boardman, 2005) This brings into question the ability of whether enterprise IT departments can meet premium levels of reliability- 99.999%- that is offered at top-level data centers, equivalent to only 5 minutes of downtime per year. (Yaple, 2006), (Marquis, 2006)

Obviously, the "traditional facilities" are being rapidly overloaded by increasing compute capacity. To get ahead of the increasing power demands, IT management needs to include data center operators, the Critical Facilities staff (CF), in the planning for new IT equipment space, power and cooling. In this way, CF can build new environments or upgrade existing facilities in anticipation of the projected IT deployments, and assure reliable operations both for day-to-day tactical operations and improve strategic planning.

Anecdotal evidence suggests that the largest enterprise IT operations, such as Yahoo!, Google and the Facebook Open Compute program (Heiliger, 2011) have already taken steps in this direction, since their core business is IT-based, and as such, the Critical Facilities staff would necessarily be involved in the strategic management of the data centers.

1.5 Problem Statement:

Information Technology Infrastructure Library (ITIL) is a management process, designed to enable the smooth deployment of Information Technology (IT) within an organization. This process is designed and utilized by IT personnel, yet facilities engineers and technicians- who manage the physical environment where the IT equipment is deployed- are not typically involved in the ITIL process, leading to inadequate coordination between IT and facilities staff. This research study was to find out if ITIL fulfills its stated function, and then determine whether the inclusion of Critical Facilities staff in the ITIL process resulted in a statistically meaningful improvement.

1.6 Research Questions:

1. Does the use of ITIL result in a significant improvement of the reliability or electrical efficiency of the data center?

2. Does the inclusion of the Critical Facilities staff in the ITIL process significantly improve the reliability or electrical efficiency of the data center?

Chapter 2: Literature Review

The Information Technology Infrastructure Library (ITIL) suite is the standard by which the Information Technology (IT) industry operates. It provides the operational and organizational framework from which effective communications are issued between stakeholders, to improve operational effectiveness and efficiencies, as well as reducing undesired operational service outages for their customers.

It is therefore useful to provide an in-depth examination of the ITIL processes with a focus on both capacity management and traditional facilities upon which all IT equipment depends. The literature review begins by providing an overview of the Information Technology Infrastructure Library (ITIL) books with regard to infrastructure planning, measurements and recommendations. The individual books are examined with regard to the definition of "infrastructure" versus the IT "environment," and capacity management recommendations.

The review then covers the capacity management issues of data centers, by the facilities staff that are responsible for delivering the power and cooling that is demanded by IT equipment, and the gaps between expectations by IT customers and delivery by facilities staff are delineated, with the ramifications of those gaps to IT-reliant businesses. The examination of existing literature revealed that there is a lack of meaningful research examining the ITIL process lack of focus on the traditional facilities of data centers, or how to effectively resolve this issue. Finally, the literature review will address recent studies on data center outages, focusing specifically on communications between stakeholders- indications that are indirect hints as to the effectiveness of ITIL within enterprise data centers.

2.1 Information Technology Infrastructure Library (ITIL) Overview

ITIL was developed in the 1980's to address the lack of standardization with regard to how IT equipment should be configured. ITIL has been through multiple revisions, the latest being "V3" issued in 2007. It is generally accepted that ITIL is the missing piece of the puzzle that enables better quality of service delivery. It facilitates saving money and time, and adds value to the organization. ITIL also spans the business and technology gap to create synergy with proven results (OGC, ITIL Continual Service Improvement, 2007 p.20).

The assumed result of ITIL is reduced total cost of ownership, smoother transition to newer technology via deployments of software and hardware, and the implementation of key performance indicators (KPIs) to assure customer expectations are met. (OGC, Continual Service Improvement, 2007, p.20). The implementation of ITIL has been effective in providing a framework with which to harness the potential horsepower that IT could deliver. Examples of these successes are prolific in contemporary culture, with such applications as downloadable music and movies (www.youtube.com), customizable radio stations (www.pandora.com), social-networking sites (Facebook and twitter), and internet new sites. Clearly, communication speeds have increased dramatically over the past decade, along with compute power. Computational speed triples every two years, a vivid example of Moore's Law, which stated that the number of processors on a chip would double every 18-24 months (Moore, 1965). From a monetary perspective, the $1000 spent in 2000, bought in 2006 a 2700% increase in computational speeds (Brill, 2007).

Yet one of the key attributes to ITIL is that the suite of books contain a variety of recommendations for 'best-practices' to manage IT service delivery, but there are no hard-and-fast rules; instead, only general suggestions are given in both process framework and metrics. The idea is that each company can tailor the suggestions to fit within their own internal framework. This lack of hard rules does provide an attractiveness for adoption by various companies that would not otherwise consider ITIL, but the generic nature of ITIL also means that key metrics are not measured exactly the same way, and processes are not consistent from company to company, resulting in some businesses being disappointed with the results (Carol, Cater-Steel, 2009).

2.2 ITIL Suite Books

ITIL addresses IT service delivery, broken into five books, title Continual Service Improvement (CSI), Service Operation (SO), Service

Design (SD), Service Strategy (SS) and Service Transition (ST) (OGC, 2007).

ITIL Continual Service Improvement (ITCSI) provides guidance for the design and development of services and Service Management processes. It covers design principles and methods for converting strategic objectives into portfolios of services and assets, with a strong emphasis on quality management, change management processes and capability improvements (OGC, Continual Service Improvement, 2007, p.6). It points out the absolutely essential need to have measureable performance measurements- KPIs- to determine whether customer needs are being met (p.20). ITCSI defines IT infrastructure as "all of the hardware, software, networks, facilities, etc., that are required to develop, test, deliver, monitor, control or support IT services" (p.200). Capacity management is briefly mentioned with regard to IT equipment (servers, data storage, networking, etc.) but no further consideration is given to the facilities which house that equipment.

An interesting aspect of the ITCSI manual is that it recognizes the difficulty of meaningful measurements in the customer service arena, and suggests the use of assessments against goals, and benchmarking against similar companies within an industry, to determine whether the studied company is performing up to market expectations (p.95-107). A key measure that will provide useful is the process maturity comparison (p.106), which is called the Capability Maturity Model Integration (CMMI), which essentially measures how diligent a company is about following their established processes.

While CMMI is used mainly for IT hardware and software deployments, the principle should be equally applicable to the use of ITIL, especially where CF staff are concerned, for planning, forecasting, problem resolution, etc. The CMMI model is the primary model I will use in my study for interactions between IT management and CF management, to determine how diligently their internal processes are followed.

ITIL Service Operation (ITSO) is the manual for best-practices for day-to-day service operations of IT applications- customer software programs and platforms (OGC ITIL Service Operation, 2007). Infrastructure is not defined in this book, though it is mentioned that IT infrastructure performance should be monitored and communicated to customers (p 190).

ITIL Service Design (ITSD) "provides guidance for the design and development of services and Service Management processes. It covers design principles and methods for converting strategic objectives into portfolios of services and assets." (OGC. ITIL Service Design, 2007)

ITSD observes that customers typically expect continuous service delivery, without regard to agreed-upon hours of operation. For them, only

24X7X365 service is acceptable. To achieve such performance requires ongoing, open communication with the customers, if one realistically hopes to achieve such levels of performance. If logically follows that those who are to deliver that performance must have access to the right information to sufficiently plan for the customers' expectations. Should the customers not want to invest sufficient resources to invest in the level of infrastructure required to attain such performance, then the customers' expectations must be managed, in that a lower investment of infrastructure will naturally result in decreased performance.

Another issue ITSD addresses, is that it is difficult to convince the business and senior management of the importance of significant investments to guarantee the expected performance. ITSD observes that the investment always comes after a failure has occurred, when it is already too late. Persuading the business and senior managers to invest before the failure occurs can only be accomplished when all IT personnel work together in producing the appropriate justifications to secure the necessary levels of investment (OGC. ITIL Service Design, 2007, p. 125).

ITSD made such observations with regard to "IT infrastructure" which it defines as IT equipment such as mainframes, servers, network equipment, data storage, software management tools, etc. The facilities of the building-power, room floor-space, cooling and physical security are labeled "environmental." (p.28)

Environmental aspects are covered in ITSD appendix E, "Environmental architectures and standards, a recommended list of standards for a particular type of facility- from an office environment to a major data center. Standards listed are aspects such as temperature and humidity ranges, access-control, network connectivity, lighting and power distribution methodologies, depending on the type of facility (p.245-248).

ITSD finally remarks upon the issue of capacity management, stating the goal is to "ensure that cost-justifiable IT capacity in all areas of IT always exist and is matched to the current and future agreed-upon needs of business, in a timely manner." (p. 79)

ITIL Service Strategy (ITSS) "is a set of specialized organizational capabilities for providing value to customers in the form of services. (OGC. ITIL Service Strategy, 2007, p.11)." Topics include development of markets, internal and external, service assets, service catalog and implementation of strategy through the service life-cycle.

ITSS takes a hard look at how to develop measurement methodology, noting measurement only becomes meaningful when it is possible to measure the actual output or dimensions of a system, function or process against a standard or desired level (p.91) The measurements that IT departments use are often meaningless to customers, because the internal measurements often bear little resemblance to what the external customer is looking for. Internal control of established companies can go so far as to slow response time to customer needs, simply due to the length of time that customers must wait, while the service provider is collecting the data for internal measurements. Indeed, they recommend an approach which promotes cooperation across the entire sphere, to attain success.

They concisely summarize the measurement issue, by stating that "Organizations have long understood the Deming principle: if you cannot measure it, you cannot manage it. Yet, despite significant investments in products and processes, many IT organizations fall short in creating a holistic service analytics capability. When combined with a disjointed translation of IT components to business processes, the results are operational models lacking in proactive or predictive capabilities." (p.197)

ITIL Service Transition (ITST) manages the transition of new or changed services into the production environment. Foci are on support modeling, workflow management, developing communication and marketing, and developing a knowledge base.

ITSD does take make an effort to discuss the question of infrastructure at length, by noting there are several layers of infrastructure, including the IT equipment types already listed elsewhere. ITSD mentions that there is an additional layer of assets includes "traditional facilities such as buildings, electricity, Heat, Ventilation, Air Conditioning (HVAC) and water supply without which it would be impossible for people, applications, and other infrastructure assets to operate. Infrastructure assets by themselves may be composed of mostly applications and other infrastructure assets. Assets viewed as applications at one level can be used as infrastructure at another. This is an important principle that allows service orientation of assets." (OGC. ITIL Service Transition, 2007 p.216)

Interestingly, ITSD then defines IT Infrastructure as "all the hardware, software, networks, facilities, etc. that are required to develop, test deliver, monitor, control or support IT services. The term IT infrastructure includes all of the Information Technology but not the associated people, processes and documentation." (p.236)

2.3 ITIL Criticisms

As the summaries above demonstrate, there is no serious consideration taken when considering the traditional facilities which deliver power and cooling. IT decisions are made in a vacuum, without sufficient consideration to upstream or downstream impacts, with data center facilities costs often excluded from overall cost calculations, and cannot tie building infrastructure to specific IT applications (Kaplan, Forrest &

Kindler, 2008). From the advent of true enterprise IT operations of the 1980's until recently, this has not been an issue. However, the escalating performance demands and dependency upon IT for most business operations has placed new pressures on the one and only area that ITIL does not address- traditional facilities which deliver the power and cooling to IT equipment. This failure of ITIL now endangers the continued IT operations of the entire industry.

A factor which worsens the situation is the lack of clear definitions in the various books, leading to confusion among IT professionals on how to proceed with change management in their field (Winniford, Conger, Erickson-Harris, 2009). This confusion naturally leads to a lack of data-center-management coordination of center's complex shared environmental and human support services, resulting extended downtime and mean time to restore (Boardman, 2005).

The lack of consideration for the traditional facilities is apparent when examining the ITIL process, yet extensive research has failed to turn up any other research, industry articles, journals or sources of any kind, which mention this issue, beyond the observation that "ITIL is an IT matter only. Business strategic, commercial and political matters, although important on an organizational scale, are not necessarily components of the ITIL implementation (Bruton, 2005)."

2.4 Ramifications of ITIL Gap

Clearly the ITIL process has had success in harnessing the full potential of IT equipment capabilities, with corresponding increases in computing and communication speeds. Yet this increase of compute power has a price- an incremental increase in the electrical power consumed. While the $1000 spent in 2006 bought 27 times more compute power than in 2000, the electrical power demanded at the plug by the newer computer was 3.4 times that of the equipment from 2000 (Brill, 2007), a 22.6% increase in power consumption year-over-year, as a result of the increased performance of IT equipment.

This represents a very real risk to the data center upon which the enterprise IT department depends. A data center is a centralized repository, either physical or virtual, for the storage, management, and dissemination of data and information organized around a particular body of knowledge or pertaining to a particular business (Godinho, 2000). Enterprise data centers are specially constructed with dedicated cooling systems (referred to as HVAC) and power distribution systems, which incorporate layers of redundancy to guarantee environmental conditions and power are uninterrupted. The data center is designed with sufficient protective systems to compensate for service interruptions, and maintain operational readiness for the Information Technology customers. For the business which relies heavily on IT, large-scale data centers are a logical choice for protecting their core business.

A key consideration in data center design is that the high levels of redundancy and reliability that are designed into the facility are planned to support a specific amount of power, which must be forecasted before construction can start. As well, while a normal commercial building has an expected life of 39 years (according to the Internal Revenue Service); a data center has an expected life of 6-15 years, with 10 years being considered typical (Rasmussen, 2011, p.2). This is largely due to the extreme level of sophistication within the facility, and the performance expectation of uninterrupted power for that entire period.

It is a given, that the typical enterprise IT department operates under a framework (ITIL) which does not factor in the power and cooling capacities that the building itself must provide. The result is that IT departments simply focus within their own circle, with little or no thought as to how much power the building must be able to provide. The warning signs of this failure to consider the traditional facilities are now manifesting, within the critical facilities industry. A variety of analyses and surveys conducted over the last five years (Brill, Aperture Research, Spears, et al) indicate that power and cooling demands are demanding at an alarming rate, with data center critical facilities engineers unable to bolster the capacities of their buildings fast enough to keep up with IT departments.

The inability of data center critical facilities personnel to keep up with demand means the support systems become vulnerable due to insufficient redundancy of power systems, with 80 percent of failures directly due to power failures (Miller, 2006). This in turn creates extreme risks for those businesses that depend heavily on IT for their core operations. In a January, 2007 study conducted by the Economist Intelligence Unit, of the U.S. National Archives and Records Administration, their findings indicated that 25 percent of companies that experienced an IT outage lasting from two to six days went bankrupt immediately. That same report also revealed that 93 percent of companies that lost their data center for 10 or more days filed for bankruptcy within a year (Gold, 2007).

A new trend appears to be emerging, however, where businesses which have an inordinate reliance upon IT reliability are actively involving critical facilities personnel in the strategic management of IT, via the ITIL process. While IT companies tend to avoid publicity on methods of operation, two public examples of this trend include the Facebook Open Compute Project and Google (Heiliger, 2011).

2.5 Conclusion

ITIL is the de facto standard for IT operations management and was developed to address the gap of how to "harness the full horsepower of IT," using standardized methods. Now that the full horsepower of IT is being utilized, the failure to include the Critical Facilities staff in the ITIL process is exposed as a fundamental flaw in this methodology. The results of this gap are widely known by the experts in the field of traditional facilities- an inability to keep up with steadily growing demands of IT equipment for cooling and power.

Some IT organizations have recognized this problem, and taken steps to mitigate the issue, though no literature has been discovered which recognizes that this gap in the ITIL methodologies represents a direct threat to the IT industry. There is no literature which explores this issue and compares performance results when the data center operators are included in the strategic planning of the data center.

The only measurement that is widely accepted across the industry to measure electrical efficiency of data centers is Power Usage Effectiveness (PUE).

There is no industry-wide measure for reliability of data centers, thus a new definition needed to be devised for the purposes of this study.

Chapter 3: Research Questions and Methodology

Information Technology Infrastructure Library, or ITIL, is considered the standard which IT departments utilize for operational management, and it is intended to provide a cost savings to IT departments by reducing operational costs and making the IT infrastructure more reliable (OGC. ITIL Continual Service Improvement, 2007, p.20).

However, a long-recognized weakness of ITIL is the inability to measure the effects of the process (OGC. ITIL Service Strategy, 2007 p.197) resulting in few studies that address its effectiveness (Persinger 2011, p.20). A detailed analysis of the ITIL suite revealed another failure in that the ITIL process lacks the inclusion of the Critical Facilities support personnel in the ITIL service delivery process. This lack of inclusion in strategic planning naturally results in insufficient cooling and/or power capacity to meet the ever-increasing demands of IT equipment. This gap in communication undermines the ITIL goals of improved reliable infrastructure, in that those personnel held directly responsible for the infrastructure are not included in the process.

This has resulted in serious consequences for the long-term viability of enterprise data centers, due to overloading of power and cooling systems, as IT equipment power usage escalates (Brill, Emerson, Spears, Crumpton, et. al.), representing significant risks to the survivability of IT-dependent corporations (Gold, 2007).

However, anecdotal evidence suggests that some enterprise IT departments are actively including the Critical Facilities personnel in the ITIL process.

The purpose of this study was to gather information from a large population of enterprise IT departments, with two distinct goals:

• Establish a measure of effectiveness of ITIL, based on generally accepted measurements of electrical efficiency and reliability, and

• Using these measures, determine whether those measures improve when Critical Facilities support staff are included in the ITIL process.

By using IT-based surveys of a large population of corporations which rely heavily upon enterprise IT-departments, quantitative analysis was used to determine whether there is a significant improvement in the overall effectiveness of the ITIL process was implemented in the data center environment, and then compare those results to those where the Critical Facilities support staff were included in that process.

The research explored the operational results of enterprise IT departments, by distributing internet-based surveys to members of large industry organizations such as The Uptime Institute and AFCOM. The surveys focused on those IT organizations (including Critical Facilities personnel), to determine:

- Does the firm use ITIL in the service management of the firm?
- What is the "maturity level" of their IT process? (OGC. ITIL Continual Service Improvement, 2007 p.106).
- Does the firm include Critical Facilities in the ITIL process, for both tactical and strategic initiative?
- If Critical Facilities personnel are involved in the ITIL process, what is the "maturity level" of the involvement of the Critical Facilities personnel?
- What is the PUE of your data center?
- How many data center outages (loss of power or cooling, which causes failure of IT equipment) have you suffered in the last five years?

"Maturity" levels are defined by internal IT Service Management Process Maturity Framework (ITSM-PMF), and are used to gauge the maturity level of ITSM processes. The assumption is that generally, organizations would wish to advance through the maturity levels, providing greater effectiveness, efficiency, and economic benefit (Persinger, 2011), and has some research appearing to confirm those assumptions (Gibson, Goldenson & Kost, 2006). The levels are as follows:

Rating	Description	Summary of Processes
0	Non-existent	Nothing present
1	Initial	Concrete evidence of development
2	Repeatable	Some process documentation but some errors likely
3	Defined	Standardized and documented
4	Managed	Monitored for compliance
5	Optimized	Processes are considered best practices through improvement

Due to the acceptance of the Capability Maturity Model Index (CMMI) within industry as a benchmarking system, the CMMI definitions were to be used to attempt to correlate how "serious" the IT department adheres to their own internal processes.

Next, to address the question of how "seriously" the Critical Facilities personnel are included in the ITIL process (if at all), the CMMI model was slightly modified to address the communications effectiveness between the Critical Facilities personnel and their IT customers. It was specifically left in a similar format to the industry-accepted CMMI model, to maintain validity and minimize additional biases of the model, as shown below:

Rating	Description	Summary of Processes
0	Non-existent	Nothing present
1	Initial	Concrete evidence of communication
2	Repeatable	Some processes involve CF staff, but some errors likely
3	Defined	Standardized and documented policies where CF staff have direct input
4	Managed	Monitored for compliance by CF staff
5	Optimized	Processes are considered best practices through improvement with direct assistance of CF staff

In addition to the CMMI measurements, the survey also included measurements generally understood and accepted by both IT and Critical Facilities personnel: electrical efficiency and reliability, as the dependent factors which measure the effectiveness of ITIL.

The measurement of electrical efficiency, commonly referred to as "PUE," is a simple calculation (total building power divided by power going to IT equipment). Therefore, this can be easily used as a measure of overall effectiveness of IT equipment deployments of all IT equipment, cooling architecture and techniques and the sharing of IT computing power amongst the IT assets.

Data center outages are not a normal occurrence, thus the number of unplanned IT service outages that have occurred in the recent past is easily remembered. The parameter for this study is referred to as Unplanned Service Outages (USO), defined as the number of unplanned failures of IT service at the respondents' primary facility due to a failure of either power or cooling delivery to the IT equipment. This isolates the measure the overall physical reliability of the data center, encompassing all cooling and power systems, the effectiveness of maintenance policies and the proper loading of power distribution nodes within the facility.

The responses of the respondents were grouped as follows:

- Those companies that do not utilize the ITIL process will immediately go to the portion of the survey which asks about efficiency and data center outages. This represented a control-group to measure the efficacy of ITIL against PUE and overall data center reliability.

- Those companies that use ITIL but do not include the Critical

Facilities (CF) staff in the ITIL process, were the first experimental group, to measure to measure the efficacy of ITIL (without CF staff) against PUE and overall data center reliability.

- Those companies that use ITIL and include the Critical Facilities staff, were the first experimental group, to see if the inclusion of CF staff made any statistically significant improvement on PUE and USO.

Independent-samples T-tests were performed to determine whether there is a statistically significant difference in the PUE and USO between those IT departments that include the Critical Facilities staff, and those that do not.

3.1 Methodology
This is a quantitative study, using an online survey (www.surveymonkey.com), to explore the relationship between reliability and electrical efficiency of corporations' enterprise data centers and how faithfully the corporation follows Information Technology Infrastructure Library (ITIL) practices, and compare those results to corporations which have included the critical facilities personnel in ITIL processes.

A quantitative survey was chosen instead of a qualitative approach, due to the fact that the respondents are geographically diverse, and work for companies where confidentiality is paramount. Thus, personal contact with respondents was automatically rejected, and no attempt was made to solicit any information that might compromise the anonymity of the respondents. The online survey vehicle was the logical choice, with invitations written to industry-specific groups on Linkedin.com.

How faithfully the corporation follows the ITIL process will be measured via the Capability Maturity Model Integration (CMMI) index, a method that is widely accepted in the industry and academia. Measurement for the level of inclusion of critical facilities personnel in the ITIL process will be performed via a slightly modified CMMI index. Logically, modification of the CMMI index to measure this parameter would establish a benchmark for something that has not been previously considered. Measurements for reliability and efficiency will be based on respondent perceptions based on specifically defined criteria for reliability and a mathematically derived result, respectively.

3.2 Development of Survey:
A fundamental weakness of the ITIL process is that it is a list of recommendations and best-practices, for companies to adapt and integrate within their own firm. That is, it is not a set of established rules, but merely suggestions on what to measure to gauge basic effectiveness. Meaningful measurement of ITIL is precluded, due to a lack of standards. Due to the admitted inability to measure real success, OGC developed the IT Service Management Process Maturity Framework (ITSM-PMF), at least allows the company to measure how genuine their internal efforts were, in the ITSM processes. Therefore, the Process Maturity Framework (PMF) will be the foundation of measuring the level of implementation not only within the IT organization, but also the critical facilities portion of the business, since it is an accepted model within industry.

Additionally, the survey attempted to use more fundamental measurements, which are generally understood and accepted by both IT and Critical Facilities personnel. It was estimated that ~70% of IT and CF personnel measure PUE [power use effectiveness] (Stansberry, 2011), and an outage is traumatic experience in enterprise IT operations, these are only two obvious indicators that both CF and IT personnel would both know intimately- the PUE of the data center, and how many outages (loss of power or cooling, which causes failure of IT equipment) have occurred over the last specific period of time.

3.3 Limitations of the Research Design:

Invitations: Since invitations were issued to voluntary industry groups via www.linkedin.com, membership to the website as well as specific membership to the industry groups must be fulfilled before the respondent received the invitation to participate. Therefore, the responses may not be a truly representative sample of the industry, merely an indication of trends.

Non-response bias: An invitation to participate was also made through the internationally-recognized trade association The Uptime Institute, but there were no responses from that avenue. The lack of response from the trade organization was verified, as the invitation went to a second survey (which was otherwise identical), yet this survey has no participants. A representative of that organization indicated this was a recurring problem, due to the membership being constantly bombarded by requests to participate in surveys. Thus there appeared no particular flaw in the survey that would deter participation from Uptime Institute members.

Survey Response Adequacy: The survey was designed to be given to a large pool of respondents. It was hoped to have at least 200 respondents, allowing the potential to explore issues such as the degree with which Critical Facilities (CF) personnel were included in the ITIL processes of the company, and explore whether there was correlation between their participation and either efficiency or reliability of their facilities. However, there were only 68 responses, of which 46 supplied at least one of the two independent variables needed for data analysis. This small response rate therefore precluded any meaningful application of the Capability Maturity Model Index (CMMI) as it is normally used in the IT venue, much less the modified CMMI to measure the inclusion of CF personnel in the ITIL process.

Confidentiality issues: the survey had to be developed in a generic format which did not require any confidential data such as contact information or the company the respondent works for. Personal experience indicates mid to high-level managers are very circumspect about the information they reveal, as the data-center management community is relatively small, requiring extreme caution about confidential corporate information. This required the survey to be completely anonymous, with no chance of backtracking to locate the respondent, and no chance of revealing confidential corporate information. This therefore precluded follow-up with any participants, so that they might expand on their views more fully.

Dependent variables: For the purposes of this survey, simple measures needed to be implemented, which were easily understood by all stakeholders of enterprise data centers; power usage effectiveness (PUE) and the number of unplanned data center outages the respondent has experienced in the last five years (USO). The former measure is a relatively 'new' measure in the industry, yet industry reports suggest that 70% of industry participants monitor this metric. Interestingly enough, only 36% of respondents were able to state the PUE of their primary facility, hinting at the actual visibility for this parameter may be significantly lower than believed.

For the purposes of this survey, simple measures needed to be implemented, which were easily understood by all stakeholders of enterprise data centers; power usage effectiveness (PUE) and the number of data center outages the respondent has experienced in the last five years. The former measure is a relatively 'new' measure in the industry, yet industry reports suggest that 70% of industry participants monitor this metric.

Other variables which might affect PUE or reliability: The building architecture may skew data for a specific respondents' PUE or reliability, but it was assumed that such biases would be minimal for the purposes of this survey, which focus more on respondent perceptions than calculated engineering numbers. Also, due to the low response rate, it was not possible to perform multivariate analysis, comparing ITIL usage, CF personnel inclusion and overall facility architecture, to determine how much of an impact the architecture had on PUE.

Unplanned service outages (USO) in a data center can occur due to a variety of causes, equipment failure, human error, environmental issues or loss of power among them. Human error aside, loss of IT service usually falls into either the loss of power, or a loss of cooling (leading to overheating and eventual failure of the equipment), and this was reflected in the survey question, without regard to whether the outage was a 'partial' outage (certain sections of the data center failed) versus a complete failure of the entire facility. This measure was also chosen as a measure of reliability, because any service outage of IT equipment due to power or cooling typically requires several days to completely recover from, and is therefore a traumatic experience to IT personnel- and Critical Facilities personnel who will have to answer to their unhappy customers after the recovery efforts are complete. Thus, the recollections of such events should be easy to recall, and fairly reliable indications of overall operational reliability.

Adequacy of Dependent Variables: From the literature review, it appears that this is the first survey taken to measure the effectiveness of ITIL against easily-determined parameters such as total number of data center outages over a period of time or PUE. It is possible that there are more appropriate variables to use, or that the definitions of the variables used in this survey were either inadequately communicated, or injected bias into the survey. Therefore, further consideration should be given to this aspect of the survey.

Multiple response bias: A last issue that may affect the overall sufficiency of the survey is that it was issued through internet forums via www.linkedin.com, without the ability to verify that a single participant did not respond multiple times. The survey also lacked the ability to preclude more than one respondent from a single company taking the survey, which may lead to skewing of data in an unpredictable manner. Knowing these flaws, the survey was therefore designed to acquire the impressions of the respondent to give indications of trends.

Self-reporting results: The quality of the survey is based on the integrity of the confidential responses from subjects. While it may be possible to cross-check responses in the survey process, the truncated survey used precluded such efforts, thus the possibility exists that some respondents did not answer honestly.

3.4 Administration of the Survey

This study concentrated on those Information Technology (IT) professionals and "traditional facilities" personnel who work in the enterprise data-center environment. The first group is comprised of those IT personnel who are concerned with information technology hardware installations, maintenance and day-to-day operations, while the latter group provides support to IT professionals and managers by providing reliable power and cooling, as well as space-planning.

Eligibility to participate in the survey was via membership in one of the www.linkedin.com industry groups, where membership is granted for vetted professionals, eliminating the tendency for non-professionals to become members of one of these groups. Thus, membership of one of the following industry groups becomes the de facto qualification to participate in the survey: the ITIL Group, Mission Critical Talent, Data Center Facility Management, Data Center Engineering, Critical Facilities and Data Center Facilities Management.

It should be noted that the survey was specifically tailored for those professionals whose employer not only operates the data center, but owns it, thus eliminating relationship problems that sometimes occur between "cloud computing" providers and their IT customers.

These two groups of subjects are represented through industry-specific groups on the professional networking website, http://www.linkedin.com. Invitations to participate were posted on the following group forums: The ITIL Group, Mission Critical Talent, Data Center Facility Management, Data Center Engineering, Critical Facilities and Data Center Facilities Management.

A pilot survey was developed, and distributed to industry colleagues to check for content validity and clarity; no obvious errors were noted. Distribution of the final survey commenced after receiving approval from the Robert Morris University IRB.

3.5 Survey Structure

The survey was structured with three specific screening questions. The first was #6, which asked, "Does your company uses the ITIL process?" If the answer was "No," then the survey will immediately move to questions #15 and 16, and the survey would be complete. In this situation, the lack of ITIL processes can be used as a control group, to track power and efficiency by those groups which do not use this process.

If the answer to question #6 was "Yes," then the survey continued on to ask the respondent their level of familiarity with ITIL via #7, and how diligently the company follows the ITIL process, using the Capability Maturity Model Index (CMMI), via questions #8.

The survey will then seek to determine whether critical facilities personnel are in the ITIL process via questions #9. If the critical facilities personnel are included in the ITIL process, questions #10 and 11 will determine if they have a different methodology they use, for tactical and strategic planning, and if so, what it is (there are some commercially available software applications available, to help with this task).

Question #12 mimics the CMMI survey question, but is slightly manipulated to measure the perception of inclusion by critical facilities personnel in the ITIL process.

Questions #13 and 14 ask how many data centers the respondent has in their company, and what the physical architecture is, of their data centers. This question may be significant, since computing-space added as an afterthought to a populated office-building tends to be less energy-efficient than a state-of-art, single-purpose data center (with no energy wasted on populated offices).

Question #15 specifically defines what an "outage" is, and then asks how many the respondent has experienced in the last five years. Data center outages are not a normal occurrence, thus the number of outages that have occurred in the near past (five years) will be easily remembered. It can be effectively used as a measure of the overall reliability of the data center, encompassing all cooling and power systems, the effectiveness of maintenance policies and the proper loading of power distribution nodes within the facility.

Question #16 defines Power Use Effectiveness (PUE), describes how to calculate it, and then asks what the PUE is of the respondents' primary data center? The measurement of PUE is a simple calculation (Total building power divided by power going to IT equipment), this can be easily used as a measure of overall effectiveness of IT equipment deployments of all IT equipment, cooling architecture and techniques and the sharing of IT computing power amongst the IT assets.

Question #1 asks where the respondent works. This is to explore whether some countries or regions of the world are more aggressive in their pursuit of ITIL processes than others.

Questions 2-4 establish how large the respondents' company is, and may indicate how important IT is to their normal operations.

Research Question	Data	Analysis
Does the use of ITIL result in a significant improvement of the reliability or electrical efficiency of the data center?	Questions 6, 14 and 16.	Independen t Samples T-test
Does the inclusion of the Critical Facilities staff in the ITIL process significantly improve the reliability or electrical efficiency of the data center?	Questions 9, 14 and 16.	Independen t Samples T-test

Chapter 4:   Data Analysis, Results and Evaluation
Initial Survey highlights:

There were 69 participants, of which 46 properly answered either one or both dependent variables, upon which to base analyses. Since the study required at least one of the dependent variables to be part of the data analysis, those participants which failed to provide either were rejected from the analysis.

Data cleanup procedures involved removing those participants which did not provide at least one of the dependent variables, as well as obvious outliers or providing improper responses to questions (for example, one respondent answered "yes" instead of providing the Power Usage Efficiency, a numerical value).

Some results from all of the participants:

50% of the respondents represented companies with annual revenues exceeding $1B, 59% of the respondents own multiple data centers. 85% of respondents identified their work as "Critical Facilities" or "Facilities Management." 48% of respondents said their Critical Facilities personnel were included in the ITIL process, but of those 20% could not answer what their level of inclusion in the process was.

Analysis:

Data analysis was performed using IBM SPSS-20.0 statistical processing software, and applying independent-samples T-tests of the responses.

4.1. Does ITIL have an impact on the electrical efficiency (PUE) of the data center?

An independent-samples t-test was conducted to compare electrical efficiency, commonly measured by Power Usage Effectiveness (PUE), against whether ITIL was used by the company. There was no significant difference in scores for those who did use ITIL (M=1.72, SD=.2) and those who did not use ITIL (M=2.11, SD=.76; t(22) = -1.805, p=.085, two-tailed). The magnitude of the differences in the means (mean difference =0.06, 95% CI: -0.85 to 0.06) was moderate (eta squared=.065). The results indicate that ITIL does not significantly improve electrical efficiency.

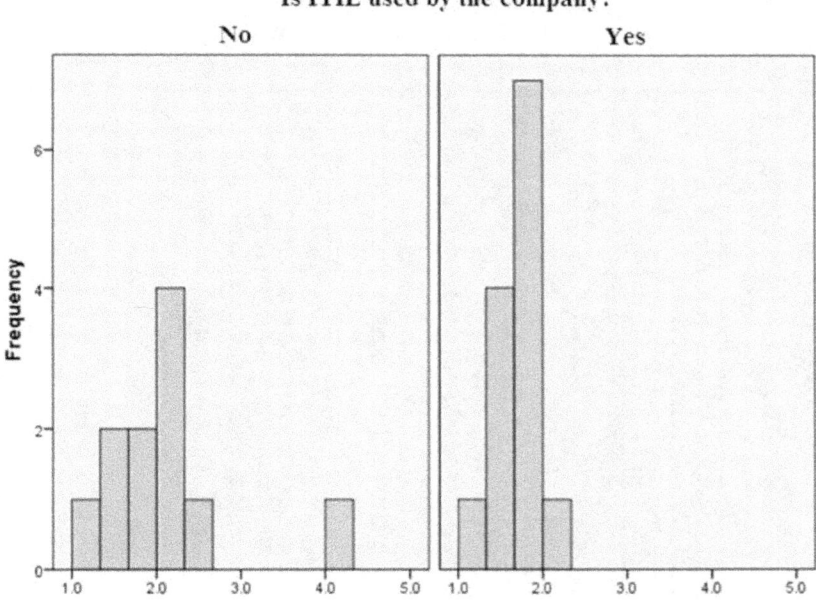

4.2 Does ITIL have an impact on the operational reliability of the data center?

An independent-samples t-test was also conducted to compare data center reliability (measure by the # of unplanned outages that have occurred in the last five years) against whether ITIL was used by the company. There was no significant difference in scores for those who did not use ITIL (M=1.24, SD=1.375) and those who did use ITIL (M=1.17, SD=1.274; t(43) = -.181, p=0.87, two-tailed). The magnitude of the differences in the means (mean difference =-0.71, 95% CI: -0.87 to 0.725) was very small (eta squared=.0007). Overall, it appears ITIL has no significant impact on data center reliability.

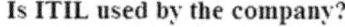

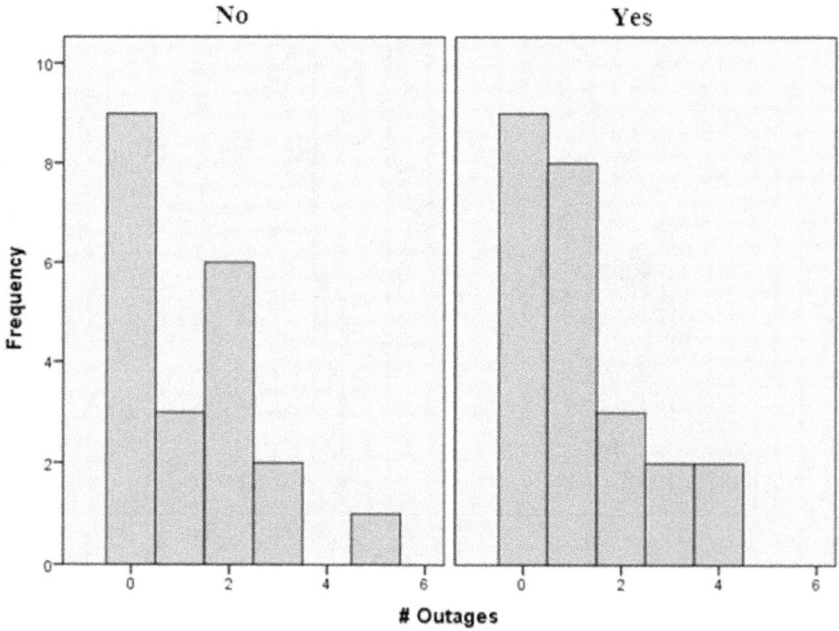

4.3 Does inclusion of the Critical Facilities personnel in the ITIL process improve operational reliability of the data center?

An independent-samples t-test compared USO (measured by the # of unplanned outages that have occurred in the last five years) against whether Critical Facilities personnel were included in the ITIL process versus those companies who did include CF personnel in the ITIL process. There was a statistically significant difference in scores for those who included CF personnel in ITIL (M=0.56, SD=0.53) and those who did not include CF personnel in the ITIL process (M=1.53, SD=1.46; t(19) = 2.35, p=0.029, two-tailed). The magnitude of the differences in the means (mean difference =0.978, 95% CI: 0.109 to 1.846) had a large effect (eta squared=.144).

Are Critical Facilities personnel included in the ITIL Process?

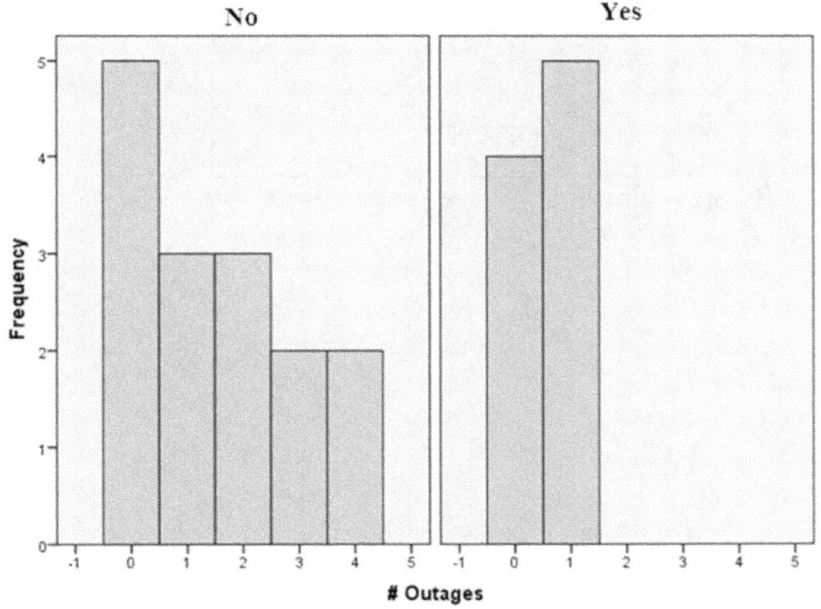

4.4 Does inclusion of the Critical Facilities personnel in the ITIL process improve electrical efficiency of the data center?

An independent-samples t-test was conducted to compare the PUE against whether Critical Facilities personnel were included in the ITIL process versus those companies who did include CF personnel in the ITIL process. There was no statistically significant difference in scores for those who included CF personnel in ITIL (M=1.725, SD=0.126) and those who did include CF personnel in the ITIL process (M=1.712, SD=0.232; t(11) = -.102, p=0.921, two-tailed). The magnitude of the differences in the means (mean difference =-.0128, 95% CI: -0.289 to 0.263) had a very small effect (eta squared=.0009

Are Critical Facilities personnel included in the ITIL Process?

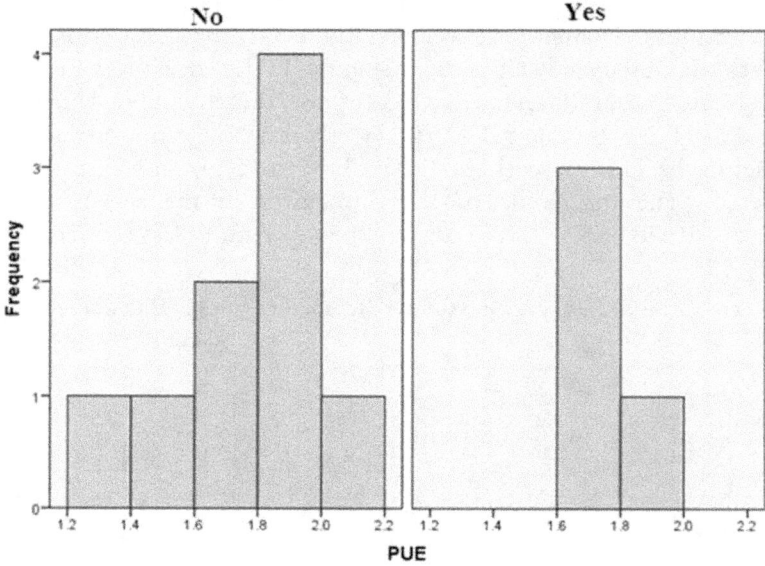

Chapter 5: Summary and Discussion
5.1 Summary

To assist the reader, this final chapter reviews the research problem and the methods used in the study. The remaining segments of this chapter summarize the results and consider their implications.

Problem Statement: Information Technology Infrastructure Library (ITIL) is management process which is designed to enable the smooth deployment of Information Technology (IT) within an organization. This process is designed and utilized by IT personnel, yet facilities engineers and technicians- who manage the physical environment where the IT equipment is deployed- are not typically involved in the ITIL process, leading to inadequate coordination between IT and facilities staff. This research study was to find out if ITIL fulfills its stated function, and then determine whether the inclusion of Critical Facilities staff in the ITIL process resulted in a statistically meaningful improvement.

The purpose of this study was to gather information from a large population of enterprise IT departments, with two distinct goals:

- Establish a measure of effectiveness of ITIL, based on generally accepted measurements of electrical efficiency and reliability, and
- Using these measures, determine whether those measures improve when Critical Facilities support staff are included in the ITIL process.

Invitations were issued on industry-specific professional networking groups via http:www.linkedin.com, to participate in web-based surveys. The surveys asked:

- Does the firm use ITIL in the service management of the firm?
- Does the firm include Critical Facilities in the ITIL process, for both tactical and strategic initiative?
- What is the PUE of your data center?
- How many data center outages (loss of power or cooling, which causes failure of IT equipment) have you suffered in the last five years?

Quantitative analysis was then used to determine whether there is a statistically significant improvement in reliability or efficiency when the ITIL process was implemented in the data center environment, and then those results were compared to those where Critical Facilities support staff were included in that process.

Summary of results

The first research question was: "Does the use of ITIL result in a significant improvement of the reliability or electrical efficiency of the data center?"

When looking at the overall effects of ITIL, the results are mixed. The overall reliability of the data center appears to become slightly better after the implementation of ITIL, though not to such a degree where it becomes statistically significant. Dispersion of responses with regard to reliability appears to be unchanged with the implementation of ITIL. Similarly, the efficiency with which power is used in the data center appears to be slightly better when ITIL is used, but again there is no statistically meaningful difference between the means. However, the implementation of ITIL appears to reduce the dispersion of responses by a factor of 3.8, indicating that while ITIL does not necessarily enhance mean levels of efficiency, it does improve consistency of energy efficiency, when ITIL is implemented. Overall, the answer to the first research question is that there appears to be no statistically significant change when ITIL is adopted by an enterprise Information Technology organization.

The next research question was "Does the inclusion of the Critical Facilities staff in the ITIL process significantly improve the reliability or electrical efficiency of the data center?" The survey data indicates that the inclusion of Critical Facilities personnel in the ITIL process has no statistically significant effect on the overall electrical efficiency of the data center, though consistency of energy efficiency is twice that of those organizations that do not include their CF staff in the ITIL process (S.D. halved).

The data indicate that the inclusion of Critical Facilities personnel in the ITIL process results in a statistically significant improvement of data center reliability, as well as increased consistency of the results (S.D. reduced by 64%).

Finally, as a follow-up to this finding, the overall reliability of the data centers were compared, between those who implemented ITIL- but failed to include the Critical Facilities in their processes- against those companies who simply didn't use ITIL at all. The mean responses would indicate that reliability is slightly worse for those companies who implemented ITIL (but failed to include Critical Facilities) than for those who didn't implement ITIL at all. However, the difference is not so strong as to be statistically significant, merely a trend. The standard deviation of their responses is indicative of a lack of consistency among the two groups- as one would expect- where no established methods of planning or control were in place.

5.2 Discussion

The data appear to show that Information Technology Infrastructure Library (ITIL) process, in and of itself, does not provide any statistically significant improvement in either power usage effectiveness (PUE) or overall data center reliability, though the dispersal of responses appears to narrow following the implementation of ITIL.

When the Critical Facilities (CF) personnel are included in the ITIL processes, however, the improvement in overall reliability improves significantly. However, PUE again does not change significantly.

If, generally, there is no meaningful difference between companies who do or do not use ITIL, yet the inclusion of CF personnel suddenly makes a difference, a logical problem comes up, "what happens when a company adopts ITIL but fails to include CF personnel in those processes?" In other words, is there a nearly-equal, but opposite reaction, if ITIL is implemented, but CF personnel are not included?

A follow-up to the third test- where the inclusion of Critical Facilities in the ITIL process appears to have a statistically significant effect of overall reliability, a final independent-samples t-test was conducted to compare the # of unplanned service outages against the respondents' company used ITIL without the involvement of Critical Facilities personnel in the ITIL process versus those companies who don't use ITIL process at all. There was no statistically significant difference in scores for those who used ITIL but didn't include CF personnel in ITIL (M=1.53, SD=1.457), and those companies who don't use the ITIL process at all (M=1.24, SD=1.375; t(34) = 0.620, p=0.54, two-tailed). The magnitude of the differences in the means (mean difference =0.295, 95% CI: -0.673 to 1.264) had a small effect (eta squared=.002). Examining this data, one finds a slight decrease in overall reliability of the data center, compared to those companies who didn't adopt the ITIL process at all.

This decrease in reliability does not appear to be statistically significant, but it does give one pause for thought. Returning to the literature review, where the ITIL manuals appear to actively ignore the CF portion of operations, the implementation of ITIL processes may distract IT professionals' attention to other activities which are more in the spotlight. An alternative possibility, is that the ITIL processes engender a false sense of security among IT professionals, resulting in less concern or attention given to the active monitoring of critical facilities.

Further research would shed light on this question.

Several weeks after the invitation to participate was issued, a flaw was discovered with regard to a specific level of the Capability Maturity Model

Index (CMMI) utilized by the ITIL process that was not included in the survey. Question #7 should have included six levels of "maturity," but level 3, "standardized and documented policies" (which should be between numbers 2 and 3 in actual survey), was omitted from the survey. Thus the true CMMI model was not properly included for the survey. However, since this question did not provide a screening function, but served merely as preparatory function before the modified CMMI index for inclusion of the Critical Facilities staff in the process, it did not affect the survey.

Turning to the ITIL process itself, it is considered the *de facto* standard by which enterprise organizations operate. In effect, ITIL is a department-wide communication model. Yet how *does* one measure whether a communication model is effective? It would seem at first blush that trying to obtain an empirical gauge of the effectiveness of a communications model defies possibility. Yet ITIL is based on the "Deming Cycle" of continuous improvement, and an important aspect of the Deming Cycle is the old business axiom, "if you can't measure it, you can't manage it." How can purveyors of the ITIL allege operational improvements, without any meaningful way to demonstrate such claims? It seemed a paradox to me.

This study was intended as a preliminary attempt to solve this problem, using terms commonly used within the industry. It is clear, however, that such measures were never intended to be applied to the ITIL framework, and therefore cannot be interpreted as a definitive appraisal of the value of ITIL.

The results of the study were necessarily constrained by the small number of respondents to the survey. Neither an in-depth analysis of the ITIL capability maturity model index (CMMI) which measures the strictness with which IT departments utilize the ITIL process nor its modified counterpart as developed for this survey, were adequately answered, to allow any meaningful comparison between diligence of the processes against efficiency or reliability of enterprise data centers. Further studies with a similar survey distributed to a large number of respondents might yield further insights and potential correlations between CF personnel inclusion in the ITIL process and efficiency or reliability.

The small number of respondents combined with the inherent biases of the survey, make it impossible to have conclusive evidence of gaps within the ITIL process. It must therefore be understood that the data merely indicates possible trends, which merit further study.

There are significant ramifications from an overall industry perspective. With a typical data center outage lasting ~90 minutes ("Understanding the cost," 2011), the respondents indicated an average of 1.53 outages in the last five years if they did not include Critical Facilities personnel in their

ITIL processes, and 0.56 outages in the last five years, if they did.

Premier IT organizations strive to achieve "five nines," or 99.999% availability. This works out to 5.25 minutes per year of unplanned outages (Yaple, Marquis), and thus calls into question the realism of such performance goals. Premier data centers advertise a design offering 99.995% availability, or 27 minutes of downtime per year. If we accept the average of 90 minutes per outage, this would infer that those companies which fail to include Critical Facilities personnel in their ITIL processes suffer an average of (1.53/5 x 90 minutes) = 27.54 minutes of downtime per year on average, which meets neither the highest IT performance goals or those of a Tier-IV data center..

However, by including Critical Facilities personnel in the ITIL process, the result is (.56/5 X 90 minutes) = 10.08 minutes of downtime per year. Clearly, this still does not match the vaunted "five nines" of high availability advertised by boutique IT organizations, but can match the performance expectations of a Tier-IV data center, by demonstrating operational sustainability that surpasses the original design intent of the architecture.

Without involving Critical Facilities in the ITIL process, it is appears equally clear that the highest expectations of IT management can never be met.

Further research into this particular aspect of ITIL, high-availability and interconnection between the two groups, I believe, is necessary.

5.3 Personal Observations and Thoughts

When the survey was developed, the hope was that there would be 200-300 respondents, with which to not only answer questions posed by the problem statement, but also allow for further research, specifically correlating a modified CMMI to measure the inclusiveness of Critical Facilities personnel in the ITIL process against PUE and USO. Unfortunately, the small number of survey respondents- while sufficient to answer the problem statement- is insufficient to use as a data-set for further research.

Having been involved in the critical facilities arena for the majority of my career, I suspected that there was a specific reason why IT departments typically did not include CF in their strategic management structure, and the results are manifest throughout the industry. However, given the discovery that certain IT-dependent corporations like Facebook were actively involving CF in their ITIL processes, I suspected that there would be a statistically significant improvement in reliability and energy efficiency, when this step was taken. I was disappointed that electrical efficiency remained unchanged, but greatly surprised at the magnitude of improvement in reliability, when CF is involved in the ITIL process structure when I discussed this approach with industry colleagues, their responses indicated that it was "common sense," in that the results were a foregone conclusion. Yet there is another axiom of the engineering world, "without data, you're just another person with an opinion."

It is obvious that the results of this research study are not the final word on the subject. On the contrary, they are merely the beginning words in a new area of research- one that I believe merits further consideration and study.

References

(2011). Understanding the cost of data center downtime: An analysis of the financial impact on infrastructure vulnerability. *Emerson Network Power,* Retrieved from http://emersonnetworkpower.com/en-US/Brands/Liebert/Documents/White Papers/data-center-uptime_24661-R05-11.pdf

Brill, K. (2007). *The invisible crisis in the data center: the economic meltdown of moore's law.* LamdaHellix.com, Retrieved from http://www.lamdahellix.com/\UserFiles\File\downloads\The_Invisible_Crisis_intheDataCenter.pdf

Bruton, N. (2005). *The ITIL experience- Has It Been Worth It?* Retrieved from http://www.rv-nrw.de/content/koop/workshops/20041206.arnw_workshop/The ITIL Experience - Hornbill.pdf

Crumpton, G. (2009). *Heat Rejection for Critical Facilities. Air Conditioning Heating & Refrigeration News,* 237(2), 37. Retrieved from Business Source Premier database.

Gibson, D., Goldenson, D., & Kost, K. (2006). *Performance results of CMMI®-based process improvement.* Manuscript submitted for publication, Software Institute, Carnegie Mellon University, Pittsburgh, PA. Retrieved from www.sei.cmu.edu/reports/06tr004.pdf

Godinho, R. (2000, September). *What is a data center?* SearchDataCenter.com, Retrieved from http://www.datacenterjournal.com/index.php?option=com_content&view=article&id=2312:what-is-a-data-center&catid=1:latest-news&Itemid=18

Gold, L. (2007, April 16). *Disaster recovery planning: How do you measure up?* Retrieved from http://www.accountingtoday.com/ato_issues/2007_7/23901-1.html

Heiliger, J. (2011, April 7). *Building efficient data centers with the open compute project.* Retrieved from http://www.facebook.com/note.php?note_id=10150144039563920

Kaplan, J., Forrest, W., & Kindler, N. (2008, July). *Revolutionizing data center efficiency.* Retrieved from http://www.mckinsey.com/clientservice/bto/pointofview/pdf/Revolutionizing_Data_Center_Efficiency.pdf

Litten, K. (2008, April 1). *Five steps to implementing ITIL.* BT. Retrieved October 15, 2011 from http://www.scribd.com/doc/3659099/Five-Steps-to-Implementing-Itil

Marquis, H. (2006, December 06). *The paradox of the 9s.* Retrieved from http://www.itsmsolutions.com/newsletters/dityvol2iss47.htm

Miller, R. (2006). *Five predictions: 9 of 10 companies face failures.* Data Center Knowledge, Retrieved from http://www.datacenterknowledge.com/archives/2006/04/03/five-predictions-9-of-10-companies-face-failures/

Moore, F. (2001). *MISSION CRITICAL FACILITIES.* Computer Technology Review, 21(1), 20. Retrieved from Business Source Premier database.

Moore, G. (1965) *"Cramming More Components Onto Integrated Circuits,"* Electronics Magazine, 19.

Mulcahy, G. (2006). *TACKLING ON-SITE POWER COSTS AT CRITICAL FACILITIES.* Power Engineering, 110(8), 38-42. Retrieved from Business Source Premier database.

Nichols, D. (2007, July 11). *ITIL doesn't matter any more (or less).* Retrieved from http://www.itsmsolutions.com/newsletters/DITYvol3iss27.htm

OGC. (2007). *ITIL Continual Service Improvement.* Norwich, England: The Stationary Office.

OGC. (2007). *ITIL Service Design.* Norwich, England: The Stationary Office.

OGC. (2007). *ITIL Service Operation.* Norwich, England: The Stationary Office.

OGC. (2007). *ITIL Service Strategy.* Norwich, England: The Stationary Office.

OGC. (2007). *ITIL Service Transition.* Norwich, England: The Stationary Office.

Organizations struggle with data center capacity management. (2006, November 23). Retrieved from http://www.emersonnetworkpower.com/en-US/Brands/Aperture/ApertureResearchInstitute/Research/Documents/ari_cap_mgmt_11_23_06.pdf

Persinger, J.. *Effectiveness of information technology infrastructure library process implementations by information technology departments within United States organizations.* Ph.D. dissertation, Indiana State University, United States -- Indiana. Retrieved September 22, 2011, from Dissertations & Theses: A&I.(Publication No. AAT 3439112).

Rasmussen, N. (2011). *Avoiding costs from oversizing data center and network room infrastructure.* Retrieved from http://www.apcmedia.com/salestools/SADE-5TNNEP_R6_EN.pdf

Spears, E. (2009, May 30). *Whitepaper: is your data center running out of power or cooling?* Retrieved from http://powerquality.eaton.com/About-Us/News-Events/whitepapers/Data-Center-Power-Cooling.asp

Stansberry, M. (2011, May 16). *Latest data center trends uncovered at uptime institute symposium 2011*. Latest Data Center Trends Uncovered at Uptime Institute Symposium 2011 | BusinessWire, Retrieved from http://www.businesswire.com/news/home/20110516005667/en/Latest-Data-Center-Trends-Uncovered-Uptime-Institute

Winniford, M., Conger, S., & Erickson-Harris, L. (2009). *Confusion in the Ranks: IT Service Management Practice and Terminology*. Information Systems Management, 26(2), 153-163. doi:10.1080/10580530902797532.

Yaple, J. (2006). *ITIL availability management: Beyond the framework*. Computer Measurement Group, Retrieved from http://www.cmg.org/measureit/issues/mit33/m_33_1.html

APPENDIX III An Analysis of Data Centers – Reducing Power Consumption and Creating Efficiencies

Eric Woodell, D.Sc.

Abstract

A key aspect of Operations Management that is often overlooked is the strategic importance of maintenance and reliability of the firm's capital assets. The result of failures can be disruptive, inconvenient, wasteful and expensive; both in terms of dollars lost due to unplanned stoppages of production, as well as rework and warranty issues that occur from degraded quality. Reliability and efficiency are paramount in any company where Information Technology is a fundamental tool of business continuity. A power outage may not only be considered an inconvenience, it can also have a significant negative financial impact on the firm. In the increasingly difficult current financial conditions, the quest for increased efficiency and energy savings is becoming a trend.

Firms must implement solutions that maintain reliability while increasing efficiency. Green IT and data center power consumption will be presented and discussed to lay the framework for analysis. The core analysis will examine the maintenance efforts of XXXXX Corporation to increase efficiency of their data center while maintaining required levels of reliability and cooling. Actual results will be compared with expected results, both in terms of monetary savings and in terms of net savings in energy consumption per computer server utilized.

Return on investment (ROI) for this project was 1,800%, with a payback period of two weeks. The documented results garnered XXXXX Corporation the 2009 Uptime Institute Award for Green IT.

Introduction

In light of globalization, the dependence on computers continuously grow in order to track the ever-increasing amount of company information and transactions, such as engineering data, records to comply with governmental laws and regulations, and the tracking of orders and shipments. The majority of enterprise servers which are capable of tracking such data are sold by Intel Corporation, observing "over the past decade, the cost of power and cooling has increased 400%, and these costs are expected to continue to rise. In some cases, power costs account for 40-50% of the total data center operation budget. To make matters worse, there is still a need to deploy more servers to support new business solutions." (Filani, Gao, He, Kumar, Nagappan, Rajappa, Shah, 2008)

Due to the increasing dependence on server and storage systems, the maintenance and reliability of the infrastructure supporting these systems has become a critical focus of firms. *93% of companies that lost their data center for 10 days or more due to a disaster filed for bankruptcy within one year of the disaster. 50% of businesses that found themselves without data management for this same time period filed for bankruptcy immediately.* (Source: National Archives and Records Administration, Washington D.C.)

Of those companies participating in the 2001 Cost of Downtime Survey, *46% said each hour of downtime would cost their companies up to $50,000, 28 percent said each hour would cost between $51,000 and $250,000, 18 percent said each hour would cost between $251,000 and $1 million, 8 percent said it would cost their companies more than $1M per hour.* (http://www.ontrack.com/library/rdr_2003_whitepaper.pdf.)

At what point does loss of data threaten the survival of a business? 40% of companies in the Cost of Downtime Survey said 72 hours, 21% said 48 hours, 15% said 24 hours, 8% said 8 hours, 9% said 4 hours, 3% said 1 hour, 4% said within the hour.? (http://www.ontrack.com/library/rdr_2003_whitepaper.pdf.)

Such concerns were confirmed by a survey in 2008 by Actionable Research, Inc. for Avocent. They polled 299 executives and IT managers in the U.S. in industries from manufacturing, high technology, retail, banking, healthcare, education and government, and among their concerns were: ("Data Center in the Hot Seat", 2008)

- *35% of companies have lost mission-critical data due to unplanned downtime.*
- *43% of companies reported they experience, on average, up to five unplanned downtime events per month. Of those, 17% experienced two to four hours of collective unplanned downtime per month, mostly due to hardware or power failures.*
- *86% of companies stated that any unplanned downtime causes a business issue, and an average of 12% lost employee productivity due to those unplanned events.*

Interestingly, they also had the following results, with regard to energy:

- 33% of companies have implemented server virtualization for energy-saving goals.
- 55% of companies can measure power usage in their data center, primarily at the Uninterruptible Power Supply (UPS) system level.
- 83% consider the ability to measure power consumption at the

entire data center level as "valuable" or "extremely valuable."

As the results demonstrate, data center reliability is critical to the very survival of firms which depend on IT for their business needs. Equally important is the inference that can be made that power consumption is a prime consideration for cost reductions, especially in the challenging financial times as we are currently experiencing.

Electrical efficiency is especially important as the amount of electricity used to power the worlds data center servers has doubled in a five-year span due mostly to an increase in demand for Internet services, such as music and video downloads, and telephony. (Ferguson, 2007) Currently data center power and cooling infrastructure worldwide utilizes more than 60,000,000 megawatt-hours per year than they actually need to power data center equipment. (Rasmussen, 2008) The increased reliability of the enterprise data center requires increased redundancy, which in turn reduces energy efficiency and increases direct energy costs. The problem then becomes how to successfully reconcile diametrically opposite goals successfully.

Data Center Energy Consumption

Data centers are notorious for having a large appetite for energy consumption. In 2006 alone, domestic company datacenters accounted for 61 billion kilowatt-hours, which is approximately 1.5 percent of all electrical usage in the U.S.(Hengst, 2008).

Data centers expend so much energy that it is estimated that by 2011, 70% of all data centers will experience operational disruptions related to energy usage and incur associated costs with diagnosing and repairing the problem (Hengst, 2008).

Companies are facing major challenges with data center energy usage. Firms are required to be as fast and efficient as possible without sacrificing data center integrity, security, and information retention. Technology, the key driving force to most datacenter functions, is creating new and innovative ways to ensure that quality isn't lost for the sake of cost-cutting.

Green IT is not simply a sweeping fad but rather a consolidated effort between domiciled and international companies coupled with the U.S. Department of Energy and other worldwide organizations to assess current data center power usages and to develop energy standards and benchmarks for data centers across the U.S. The Green IT movement is rapidly becoming a way of life for companies and regulations are being set forth requiring companies to conform virtually immediately to reduce energy consumption. Green IT is here to stay.

Complying with stringent regulations is forcing companies to find new and innovative ways to reduce energy consumption without having to build brand new state-of-the-art data centers. Companies are relying on guidelines drafted by organization such as The Green Grid, a collection of IT companies with the mission and purpose to raise the energy efficiency of data centers, to meet current energy consumption requirements (Hengst, 2008). Simple techniques such as benchmarking performance, switching to energy-efficient lighting, improving archiving procedures and optimizing air-conditioner settings are just a few tactics than can reduce energy.

Other more in-depth energy saving methods such as consolidation of servers with virtualization, converting to intelligent or single-instance storage, and re-configuring server rooms are also proving to be effective tools for data centers to implement (Hengst, 2008).

A tidal wave of solutions are available for data centers to reduce energy consumption. These solutions seem attractive and cost-effective on paper, but many organizations are finding it difficult to know exactly where to begin. Luckily, IT industry giants such as IBM are available to help companies struggling to begin the "green conversion". IBM is aggressively marketing green data center solutions and servers to data center managers striving to comply with current regulations and reduce costs as well (Higdon, 2007).

Green IT and the Uptime Institute

To address the concerns of managing the maintenance and reliability of enterprise data centers, the Uptime Institute was founded in 1993. Its purpose is to perform research focusing on data center facilities, the IT and facilities interface, and how both functions affect the cost, reliability and energy consumption of computing.

Institute Best Practices arise out of work with the practitioners who actually run data centers, comparative cost and uptime statistics, consulting work, surveys, and internally funded efforts.

The Institute has pioneered and continues to develop numerous innovations that subsequently have become industry standards, including hot/cold aisles, site infrastructure energy overhead metrics and measurements, cost modeling, the dual-power specification, and the Tier Classification system for rating the concurrent maintainability and fault tolerance of data center facilities. (Uptime Institute.).

The Uptime Institute has researched and disseminated a list of best practices for support systems within the enterprise data center, to maintain continuous availability. Such practices include the utilization of Information Technology System Management (ITSM) software and network support for physical components (routers, switches, hubs), electrical systems, including battery-backed UPS systems, multiple redundant computer-room air-conditioners (CRACs), fire-suppression systems, smoke-detectors and video surveillance for security.

Uptime Institute recommends that each system have specific levels of redundancy and flexibility, to assure functionality in the event of a catastrophic event and allow concurrent maintenance, while production equipment is still online.

In 2009 the Uptime Institute established the 'Green IT Award" to reward companies who are taking on green initiatives and reducing their environmental impact. This reward is limited to Fortune 500 and InformationWeek 500 companies that are committed to increasing energy efficiency and reducing their environmental impact. In 2009, the example company received the award for the strides they have made towards becoming a Green IT Corporation as well as the commitment to continue the initiative.

Introduction of the Example Corporation

XXXXXXXXX is the service division of XXXXX Corporation tasked with handling all Information Technology issues for other XXXXX Corporation divisions, including information storage and access, software application development and maintenance, operations software necessary for day-to-day operations, and the storage of production and engineering data, in accordance with local, state and federal laws, as well as regulations specified by the Food and Drug Administration. XXXXXXXXX is also responsible for all E-commerce related applications. Total revenue of XXXXX North America last year was $26 billion. All of the revenue generating transactions are stored and routed through the XXXXX Corporation data center; the estimated cost of a 30-second power interruption in the data center is in excess of $70 million. With the data center providing the crucial support for the entire corporation, system architecture is configured as "2N+2", making it one of the most redundant and resilient support systems.

The case-study of XXXXX Corporation examines the results of maintenance efforts to increase efficiency of their data center, while maintaining required levels of reliability and cooling, and compares to the standards set forth by the Uptime Institute best-practices recommendations.

In January, 2007, XXXXX Corporation unveiled their Green IT program, designed to solicit responses throughout the company, for ideas to help improve the "greenness" of XXXXX Corporation. These included earth berming, heat-reflective material on the roof, geothermal cooling and solar power.

1. Earth berming involves the application of dirt against the sides of the building, with the inherent cooling effect, to absorb heat

through the walls via convection. This idea was quickly rejected, after an engineering analysis determined that the structure of the building could not support tons of earth pushing on the walls, causing the building to collapse. (figure. 1)

2. Heat-reflective material on the roof was proposed to reflect sunlight away from the building during summer months in order to help keep the building cool. This idea was rejected, because while the capital costs were moderate, there was no empirical data to indicate this would have any significant impact on cooling efficiency within the data center. (figure 2)

3. Geothermal cooling is the use of long tunnels, placed more than 20 feet below the ground, to cool the air. Air would be drawn through these tunnels, be cooled by the air, and then re-directed into the data center. This idea was rejected as unfeasible, given the rocky nature of the terrain, the high capital costs involved, and the uncertainty of efficiency of the system (figure 3).

4. Solar power was rejected because it would have a negative Net Present Value. Given the uncertain weather of the geographic area, the lack of intense sunlight even during clear summer days, the excessive capital expenditures required to install solar power panels, and the extreme number of panels required to yield any noticeable energy savings, this approach, after lengthy engineering studies, was deemed unfeasible (figure 4).

After these analyses were complete, it was clear that the XXXXX Data Center needed to "go back to basics," looking for ways to increase the efficiency of existing infrastructure with minimal capital outlay and maximum effectiveness. Initial problem solving began by doing a thorough examination of data center architecture, including power distribution systems, HVAC (Heating, Ventilation, and Air Conditioning), power and fiber routing under the floor (which may impede cooling airflow to where it is needed), and configuration of the CRAC (Computer Room Air Conditioner) units. After the survey was complete, the results were compared with the best-practices provided by the Uptime Institute and corrective actions were implemented in a sequential fashion, to correlate actual results against expectations.

Initial Analysis at XXXXX

According to the Uptime Institute, the Site Infrastructure Energy Efficiency Ratio (PUE) is a simple way to determine how well a data center is doing, in managing the efficiency of their data center's site infrastructure systems. Simply stated, it is a measure of power "in" to the data center as measured at the utility electric meter divided by the conditioned power "out" to run the IT equipment for computing. The difference between output and input are transformation losses, Uninterruptible Power Supply (battery-backed "UPS" systems) and cooling equipment inefficiencies, and operational choices that are under the full control of the Mission Critical Facilities function.

According to Kenneth Brill of the Uptime Institute, the typical data center has a PUE of 2.5. This means that for every 2.5 watts "in" at the data center utility meter, only 1 watt is delivered "out" to the IT critical load. The best ratio possible is 1.6, due to a variety of physical factors and inefficiencies (beyond the scope of this analysis). (Brill, 2007) This ratio is possible in data centers with minimal-to-medium redundancies. The best possible ratio for data centers that offers maximum protection and reliability is approximately 2.0. According to Brill, for many large data centers (greater than 30,000 square feet) improving the PUE from 2.5 to 2.0 will save close to $1M annually, with no adverse impact on reliability and may actually improve reliability.

Initial measurements of the XXXXX Corporation data center indicated a PUE of 2.66, indicating that there were ample opportunities to improve efficiency.

Strategic Move From Physical to Virtual Servers:

It has been widely accepted within the IT industry, that moving from (for example) 20 physical servers to 20 virtual servers within a single computer (purpose-built to run virtual servers) will generally yield a 15-20% savings in overall energy (Niles, 2008). Virtual servers are applications that function in the same way that a physical computer server does, but it has the key feature in that it can be started or stopped on any server that supports Virtual Machines ("VM-ware" – Virtual Memory Software). In July of 2007, the IT Operations managers decided to install a limited number of virtual machines as an experiment to study the effectiveness of this approach. By July of 2008, in an aggressive move to pursue a Green IT strategy, XXXXX Corporation implemented a new policy to force the use of VM-ware on all server "owners," and would install physical servers only when the server owner could provide a solid justification for it.

Facilities staff assumed that the increasing number of servers being used was partially responsible for the excessive energy consumption of the data center, even though the use of VM-ware was becoming more common practice. A review of the number of servers- both virtual and physical- was conducted, to attempt to find a correlation between power consumption and the number of servers. The results were that, while the number of physical servers remained effectively unchanged, IT demands had increased 50% in just 27 months, and were met by using virtual servers (fig. 6).

It was determined that the net power consumption levels were not directly impacted by server counts. In comparing power consumption to physical server counts, there was no direct correlation (compare figures 6 & 7 over the same time-frame), and it was concluded that the total number of servers was not relevant to the excessive power consumption by the data center.

Bypass Airflow Causing Overheating:

Cooling to the individual servers is a critical nature of DC operations; without it servers will quickly overheat and shut down. In order to help prevent this, the entire production area of a DC is built on a raised floor. Hot air is drawn in at the top of the room by the CRAC units, chilled via air-cooled air-conditioning systems and the cold air discharged under the floor. The floor has openings at strategic locations, so that cooling air is discharged where it can be used to cool servers (see figure 5). For maximum performance and reliability, computer manufacturers recommend a maximum temperature of less than 77^0F with a rate of change not exceeding 9^0F per hour, due to limitations on how resistant to thermal expansion/contraction computer circuit boards are. Maximum air inlet temperature should not exceed 75^0F. (Grill, Strong, Sullivan, 2006) In the XXXXX Corporation data center, a review of temperature profiles revealed the presence of hot "zones," with approximately 20% of computer servers in production having air inlet temperatures in excess of 75^0 FF.

"Bypass airflow" is defined as conditioned air that is short-cycled or not getting directly into the computer equipment. This air escapes through cable cutouts, holes under cabinets, or misplaced perforated tiles (Sullivan, Strong and Brill, 2006). (figure 5)

Subsequent investigation of the XXXXX Corporation data center yielded the detection of several hot-spots and demonstrated that the data center was suffering significant bypass airflow problems. Approximately 7.5% of the production floor had openings in areas where there were no heat-loads, allowing cold air to escape where it was not being used. This caused a lack of pressure within the entire raised-floor cooling architecture, preventing cold air from getting to where it was most needed.

Subsequent investigation of the XXXXX Corporation data center yielded the detection of several hot-spots and demonstrated that the data center was suffering significant bypass airflow problems. Approximately 7.5% of the production floor had openings in areas where there were no heat-loads, allowing cold air to escape where it was not being used. This caused a lack of pressure within the entire raised-floor cooling architecture, preventing cold air from getting to where it was most needed.

As the XXXXX Data Center is a facility requiring continuous

availability, disconnecting computer cables to install seals under computer servers was not an option. After additional analysis the decision was made to install fire-resistant silicon-based foam into custom frames and all holes that were not directly cooling IT production equipment were sealed off. While no initial under-floor pressure measurements were performed, upon completion of after this step was completed there was a noticeable increase in under-floor air pressure and in the presence of cooling air at IT loads.

Computer Room Air Conditioner Redundancies:

After the bypass airflow issue was resolved, focus next turned to the CRAC units, which supply the cold air under the floor. The power consumed by critical loads was compared to the cooling capacity of the CRAC units. Since air-conditioning units are rated in tons of capacity, the electrical load of the data center (in kilowatts) was converted to tons, and compared to the total cooling capacity of air-conditioning units that were running in the data center. At the beginning of the project, the raised-floor production areas were using 490 kW, which translates into 139.4 tons of air-conditioner cooling capacity. Total CRAC unit capacity at the time the initial readings were taken indicated 366 tons available, thus, the CRAC units were running at 139.4/366=38% efficiency.

Equipment such as air handlers, chillers, cooling towers, condensers, pumps and dry coolers consume some power while performing their cooling function (that is, some of their input power is dispersed as heat instead of contributing to the mechanical work of cooling). In fact, the inefficiency (waste heat) of cooling equipment typically greatly exceeds the inefficiency (waste heat) of power equipment. When cooling equipment is doubled for redundancy or when the equipment is operated well below its rated power, efficiency falls dramatically. Therefore, an increase in efficiency of the cooling equipment benefits overall system efficiency. (Rasmussen, 2008)

Detailed calculations indicated that the inefficiencies due to operating twice the necessary CRAC units, was largely responsible for the data center PUE. After consulting with top managers of XXXX Business Services, an outside consulting firm (AdaptivCool, Inc.) was invited to perform a detailed computer fluid-dynamics study of the XXXXX Corporation Data Center. Their preliminary findings (example figures 8, 9), confirmed that too many CRAC units were deployed, causing disruptions in airflow. Eight CRAC units (of seventeen units in use) were deemed to be hindering cooling effectiveness.

AdaptivCool also discovered significant airflow bypass issues occurring between the upper and lower floors of the data center. Immediate actions were taken to seal the path for airflow bypass between floors. After this was complete, CRAC units were sequentially secured with careful monitoring of temperatures to verify that all servers remained within environmental specifications. At the end of this process, a subsequent review of conditions indicated that the environment was stable.

Results

☐ The Power Usage Efficiency is a relatively simple method to determine how well the data center is doing in terms of efficiency. Overall PUE went from 2.66 to 2.3, however, the incremental results using this approach were too small to be useful, in determining the effectiveness of each action performed. The most meaningful analysis available in this case is the determination of power consumption per server installed within the XXXXX Corporation Data Center, and this proved very instructive (figure 11).

☐ From March of 2008 to August 2008, one of the capital projects in progress was the installation of CRAC units that were "air-cooled," meaning that they use R-12 refrigerant for cooling, to replace water-cooled units. While this reduced data center risk of a power outage due to a water leak under the raised floor, water is an inherently more efficient cooling medium than R-12, due to the evaporation properties of water-cooled air conditioners. This explains the slow rise of power consumption during this time period.

☐ From August 2008 to December 2008, the project to resolve the airflow bypass problems was underway, and as the graph shows, some efficiency was captured, yielding an efficiency increase of approximately 13.33%.

☐ Excess cooling capacity was secured January 6, 2009, resulting in a significant decrease in the amount of energy consumed per server, approximately 24%.

☐ Cumulative results took energy consumption per server from 800 watts of energy consumed for each server, down to 350 watts, a 56% decrease.

☐ Cumulative monetary savings due to these simple measures equaled $130,000 per year.

Future Plans

While the improvement in PUE was encouraging, the fact remains that there are still opportunities to increase efficiency further. Following the shut-down of redundant CRAC units, another analysis was performed of the XXXXX Corporation data center infrastructure. The only significant discovery was that the building operates with two distinct power "loops," so that if one loop fails, all servers and network equipment can maintain functionality using the opposite loop. However, in each loop, there are three UPS modules, each rated for 500 kilowatts of power that operate in parallel.

One loop is loaded with 230 kilowatts of power (output to the production floor), and draws an input of 260 kilowatts. The difference of input versus output is due to the electrical energy used to keep each UPS module functioning, or 20 kilowatts for each module, and is an expected difference. The other loop is loaded with 150 kilowatts (output to the production floor), yet its input power was also 260 kilowatts.

The extra power consumed by the UPS loop with lighter loading, is due to "reactive power," (which is commonly known in industrial power distribution systems), which is affected by the types of electrical loads placed upon them. However, reactive loads also can exist in very rare instances, when two or more electrical generators are sharing a load. If the loading becomes too small, relative to the designed electrical load for the generators running in parallel, the generator loads become unbalanced, and the generators begin to fight each other for frequency and voltage control. When three or more lightly loaded generators are operating in parallel, the instability increases due to reactive power as a square of the number of generators operating in parallel.

In the XXXXX Corporation data center, the lighter-loaded UPS modules act as generators, with many of the same characteristics of more common generators that are found in industrial and agricultural applications. With them carrying only 10% of the rated load, they are now becoming unstable in synchronization and voltage control, and start to "fight" each other for control of the system they are supplying. In such situations, the loads are referred to as "phantom" loads, and consume power from the input side, but don't appear as output to the IT equipment, because the internal resistances of the various pieces of IT equipment are what dictate "real" load.

To correct the presence of reactive loads between the three AC generators, there are two options available. The first is to increase real electrical load to a level above the minimum performance envelope of all three UPS modules, so that they revert to their normal energy consumption (as witnessed by the UPS with higher loads). Unfortunately, this option cannot be implemented, as it is a function of production equipment that XXXXX Corporation has purchased. The second option is to shut down one of the UPS modules from each loop, thereby increasing the electrical loading on the two other UPS modules above the lower-limit of design-load envelope, triggering the "overhead" power consumed by the UPS modules to decrease by 33% for each side, further increasing electrical efficiency of the data center.

Current calculations indicate that this step, once completed, will save XXXXX Corporation an additional 63,000 kilowatt-hours each month, translating into extra savings of $75,000 annually, above the savings already being realized. This final step will decrease Power Usage Efficiency further, for a net result of 2.08.

It is our opinion that further reductions in energy consumption, without a highly automated monitoring and control system, are not possible while still maintaining the maximum reliability of the XXXXX Corporation Data Center.

One distinct challenge of the XXXXX Corporation data center, going forward from their current situation, is that a data center, by nature, is a dynamic environment. IT personnel are constantly installing and removing production equipment, which change overall power requirements to the data center, over time. Even more crucial, is that the cooling system is a fluidic medium; thus, any changes in the data center can cause eddies and swirls of airflow, like a river, changing the effectiveness of the cooling system. These dynamic properties of a data center, will require that a high level of vigilance be maintained by Facilities Team personnel, to monitor conditions and take corrective actions as needed, to preclude production interruptions going into the future.

Conclusions

The task of continuing maintenance to assure reliability of the firm's physical assets is a hidden but crucial task of Operations Management. A lapse in the continuing maintenance coverage can result in significant financial repercussions for the firm or employees, up to and including death of the employee *or* the firm.

In contemporary businesses, the increasing dependency on computers, networks and storage systems to support globalization offers a double-edge sword. On the one hand, it offers the ability to manage and interpret immeasurable amounts of raw data, turning it into useful information, as well as allowing unprecedented levels of communication and efficiency. Yet, on the other hand, that very reliance on IT places a heavier burden on the support systems for data centers, making it the new Achilles Heel for modern businesses. This weakness must be recognized and accounted for if the -financial impact of improper maintenance of the data center is taken into account.

Figures:

Figure1:

Figure 2:

Figure 3:

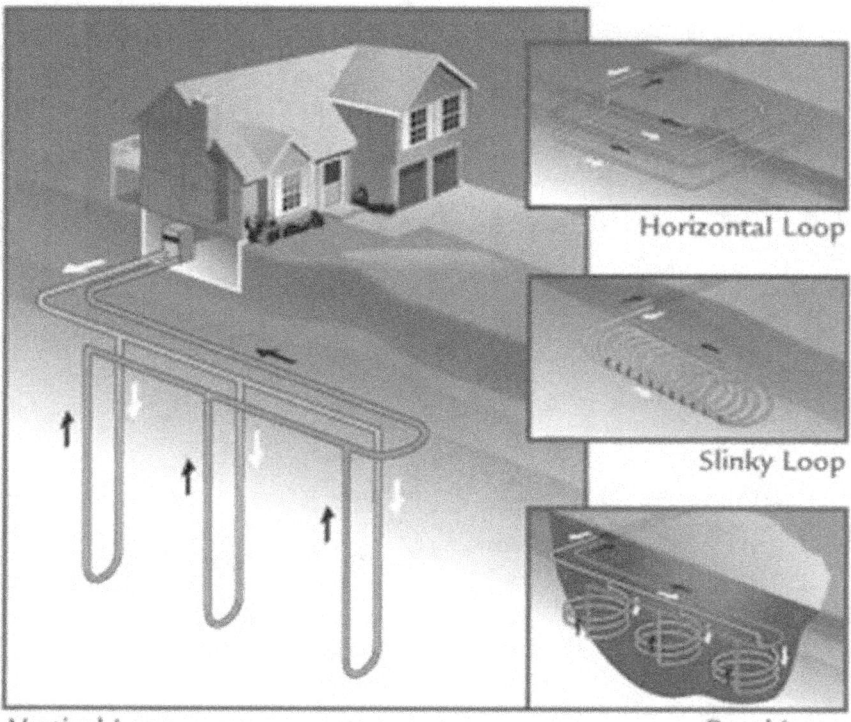

Horizontal Loop

Slinky Loop

Vertical Loop

Pond Loop

Figure 4:

Figure 5:

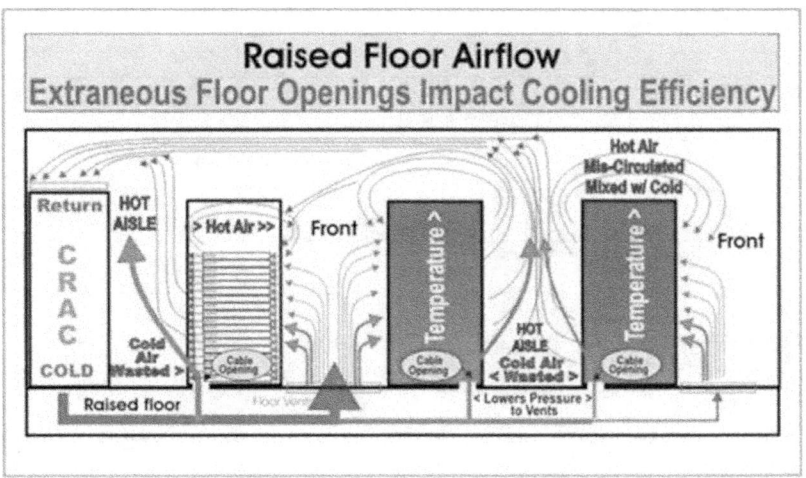

Figure 6:

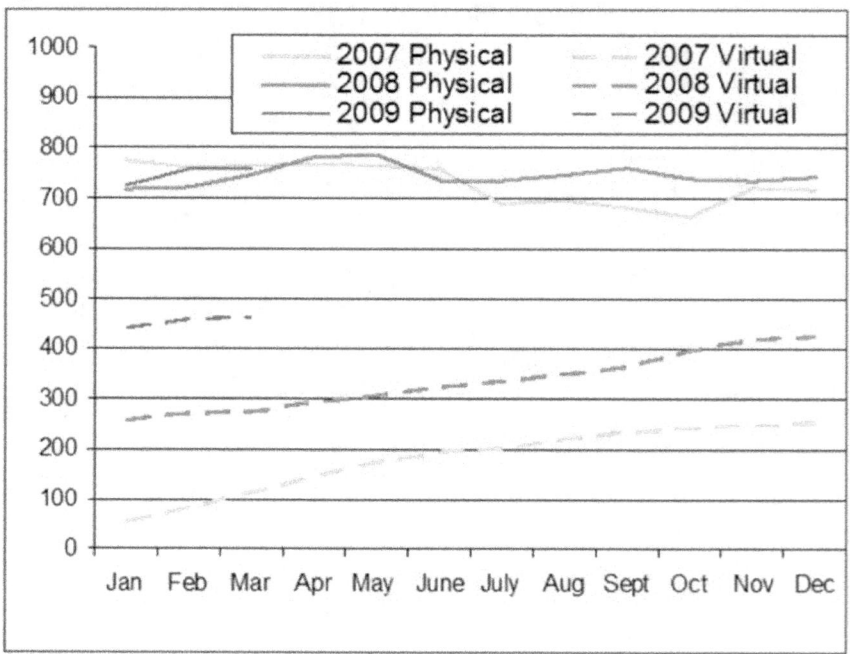

Figure 7:

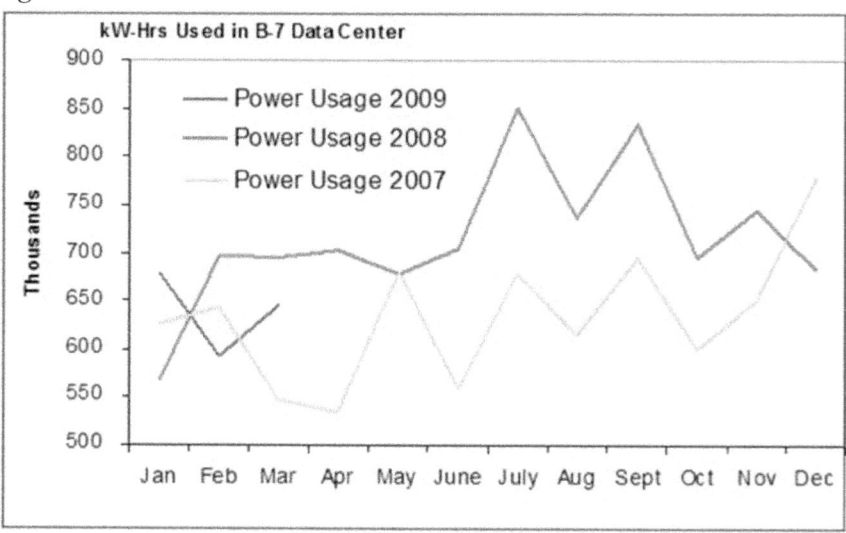

Figure 8:

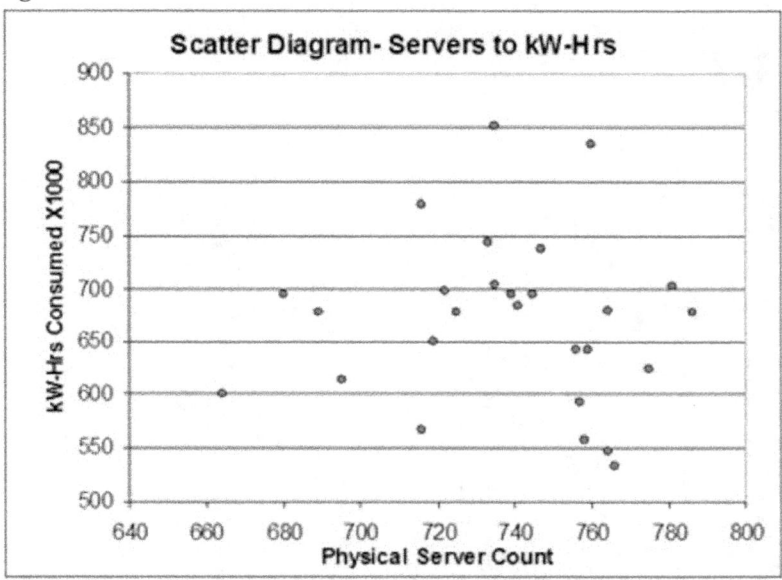

Figure 9:

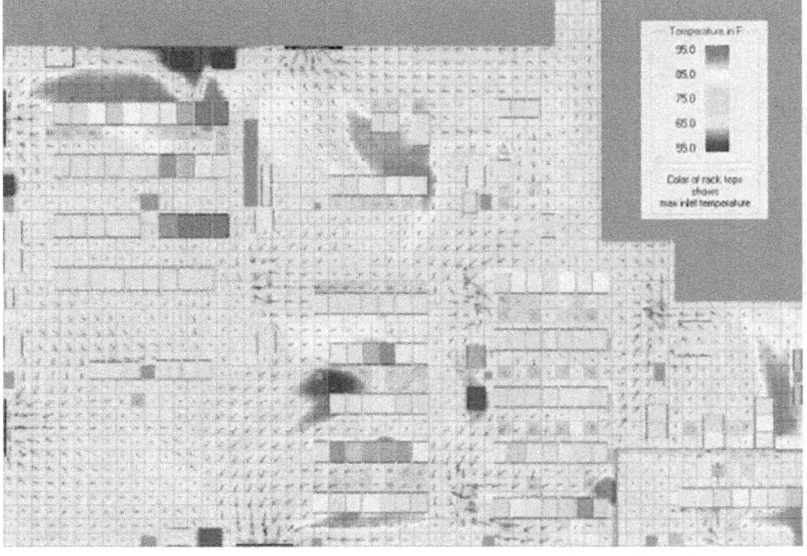

Figure 10:

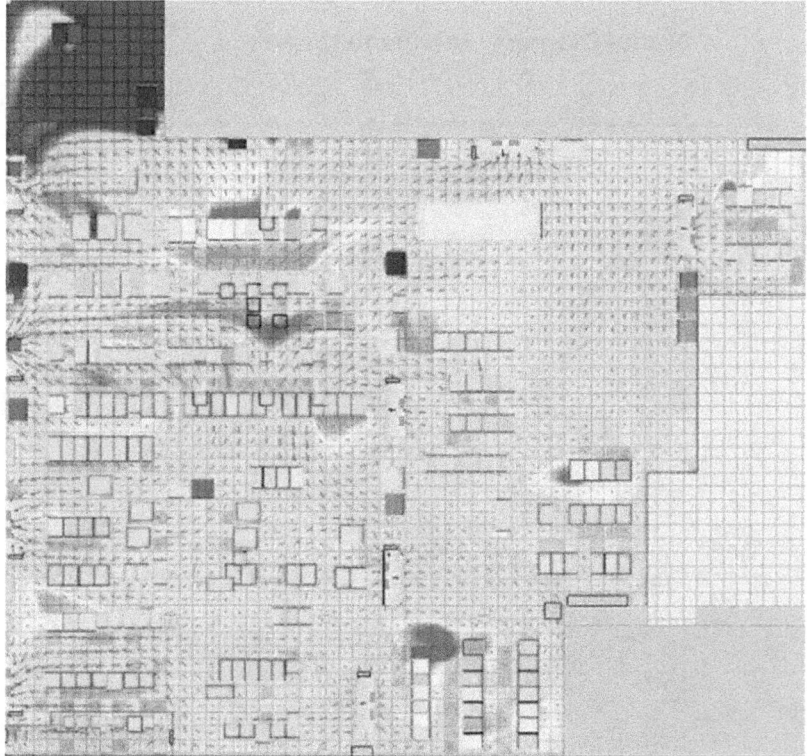

Figure 11:

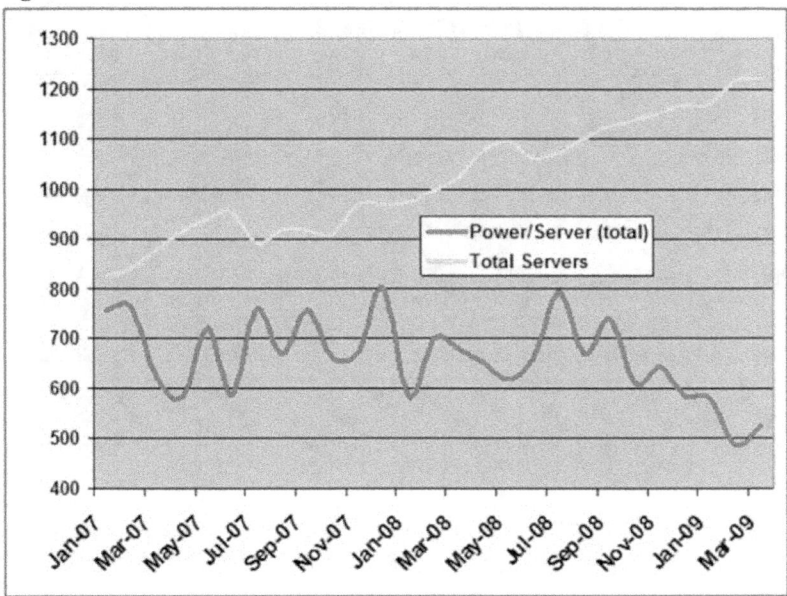

Figure 12:

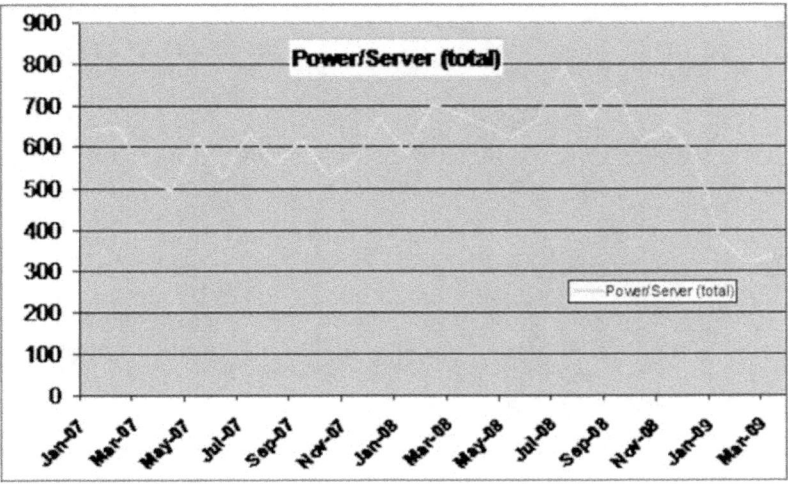

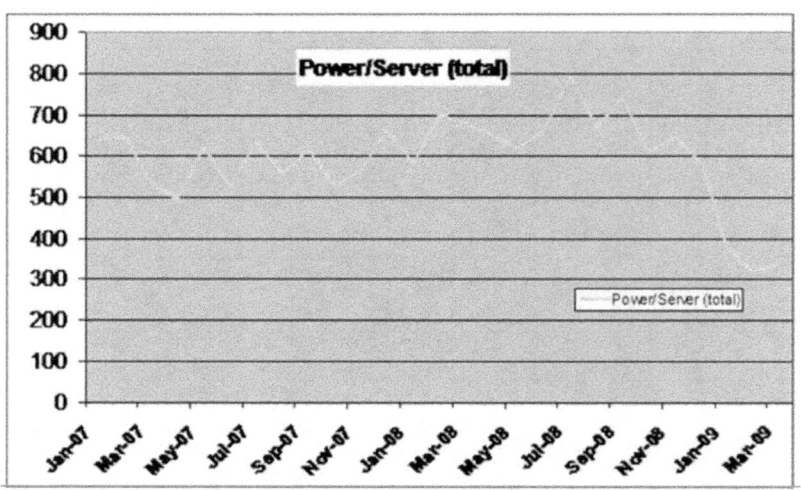

References

Uptime Institute Homepage. Retrieved April 21, 2009 from http://uptimeinstitute.org/content/view/57/81

Green Data Center. Retrieved April 21, 2009 from http://searchdatacenter.techtarget.com/sDefinition/0 ,,sid80_gci1178582,00.html

Schulz, Greg. (2009). "Enabling a Green and Virtual Data Center". *GreenerComputing.*

Retrieved April 21, 2009 from http://www.greenercomputing.com/feature/2009/02/18/enabling-green-virtual-data-center

Uptime Institute Honors 2009 Global Green 100 for Corporate Leadership in IT Energy

Efficiency. (2009). *Uptime Institue.* Retrieved on April 21, 2009 from http://uptimeinstitute.org/content/view/306/286/

Hengst, Amy. (2008). "10 Ways to Improve Power Performance in Your Datacenter", *IT Management".* Retrieved April 21, 2009 from http://www.itmanagement.com/features/improve-power-performance-datacenter-100407/

Higdon, Jim. (2007). "Reducing Datacenter Consumption", *IT Management,* Retrieved April 21, 2009 from http://www.itmanagement.com/features/datacenter-power-consumption-061907/

Niles, S. 2008. "Virtualization- Optimized Power Cooling". Retrieved April 7, 2009 from http://www.apcmedia.com/salestools/SNIS-7AULCP_R1_EN.pdf

Turner, P. (2002). "Changing Cooling Requirements Leave Many Data Centers at Risk". Engineered Systems Retrieved April 14, 2009 from http://www.uptimeinstitute.org/cgi-bin/admin2/admin.pl?admin=view_whitepapers

Antonopoulos, Andreas M. (2006). "Green data centers help the bottom line" *NetworkWorld.* Retrieved April 14, 2009 from www.networkworld.com/newsletters/datacenter/2006/0227 datacenter1.html

Sullivan, R.F., Strong, L., Brill, K. (2007). "Reducing Bypass Airflow Is Essential for Eliminating Computer Room Hot Spots" *Uptime Institute White Paper.* Retrieved April 14, 2009

Brill, K. (2007). "Data Center Energy Efficiency and Productivity" *Uptime Institute White Paper.* Retrieved April 14, 2009 from http://www.uptimeinstitute.org/symp_pdf/(TUI3004C)DataCenterEnergy Efficiency.pdf

Brill, K. (2009). "Revolutionizing Data Center Efficiency (Update)". *Uptime Institute White Paper.* Retrieved April 14, 2009 from http://www.uptimeinstitute.org/cgi-bin/admin2/admin.pl?admin=view_whitepapers

Briody, D. (2006). "10 simple steps to a green data center". *CIO Insight.* Retrieved April 14, 2009 from www.cioinsight.com/article2/0,1540,2020883,00.asp

Crosby, C. 2007. "A common sense guide to 'green' planning & design for data centers." *The Data Center Journal.* April 14, 2009 from http://datacenterjournal.com/index.php?option=com_content&task=view&id=978&Itemid=134

Doty, S. (2006). "Energy efficiency in computer data centers". *Energy Engineering* 103(5):50 – 76. Retrieved on April 14, 2009 from http://www.csu.org/environment/conservation_bus/energy/library/18746.pdf

Intel. (2006). "Increasing Data Center Density While Driving Down Power and Cooling Costs". *Intel White Paper.* Retrieved on April 14, 2009 from www.intel.com/business/bss/infrastructure/enterprise/power_thermal.pdf.

Rasmussen, N. (2006). "Implementing Energy Efficient Data Centers." *APC White Paper.* Retrieved on April 14, 2009 from http://www.apcmedia.com/salestools/NRAN-6LXSHX_R0_EN.pdf

Rasmussen, N. (2007). "Calculating Total Cooling Requirements for Data Centers.". APC White Paper. Retrieved April 14, 2009 from http://www.apcmedia.com/salestools/NRAN-5TE6HE_R2_EN.pdf

Rasmussen, N. (2008). "An Improved Architecture for High-Efficiency, High-Density Data Centers". *APC White Paper.* Retrieved on April 14, 2009 from http://www.apcmedia.com/salestools/NRAN-6V5QAA_R0_EN.pdf

Gowri, K. 2004. "Green building rating systems: an overview." *ASHRAE Journal.* Retrieved on April 14, 2009 from http://www.energycodes.gov/implement/pdfs/Sustainability.pdf

U.S. Environmental Protection Agency ENERGY STAR® Program. (2007). "Report to Congress on Server and Data Center Energy Efficiency Public Law 109-431". Retrieved on April 14, 2009 from http://www.energystar.gov/ia/partners/prod_development/downloads/EPA_Datacenter_Report_Congress_Final1.pdf

Filani , David., Gao, Sam., He, Jackson., Kumar, Anil., Nagappan, Ram., Rajappa, Murali., Shah, Pinkesh.(2008). "Dynamic Data Center Power Management: Trends, Issues, and Solutions,". *TechRepublic.* Retrieved on April 17, 2009 from http://whitepapers.techrepublic.com.com/abstract.aspx?docid=353996

"Data Center in the Hot Seat". (2008). *Avocent Webpage*. Retrieved on April 17, 2009 from http://www.avocentinfo.com/10135/index.php?d=domorewithless

Ferguson, S. (2007). "Data Center Power Consumption on the Rise, Report Shows". *eWeek*. Retrieved on April 17, 2009 from http://www.eweek.com/c/a/IT-Infrastructure/Data-Center-Power-Consumption-on-the-Rise-Report-Shows/

APPENDIX IV: SKILLS COMPETENCIES DEVELOPMENT PROGRAM (SCDP)

SCDP stands for Skills Competencies Development Program. SCDP is a program intended as a means for XXXXX Data Center personnel to progress from being newly hired, being qualified to be part of the Duty Engineer rotation, and to advance beyond that to a higher level of competency and responsibility. To this end, the SCDP team has the following goals:

    1. Define the different levels of achievement, skill and knowledge necessary to maintain a data center.

    2. Determine the path towards these different levels to provide personnel with a clear vision and direction for professional growth.

    3. Incorporate training and personnel development into the pathway for this professional growth to occur,

    4. To outline a means to adequately document and track personnel growth, skills, abilities, level of knowledge and experiences to accurately determine what level personnel have attained as they progress in this program.

SUPPORT ASSETS FACILITIES SKILLS COMPETENCIES DEVELOPMENT PROGRAM

Statement of Purpose:

The purpose of the Support Assets Critical Operations Facilities Skills Competencies Development Program is to maintain the highest degree of Data Center Availability, by ensuring that individuals responsible for the support of the Critical Equipment and Systems are properly trained as Certified Duty Engineers.

Initial Candidate Evaluation:

Support Assets Critical Operations Management needs to decide where to start a candidate, when outlining the initial certification flowpath. This will be based on an initial evaluation of the candidate's skills, knowledge, and background, to establish a best path for candidate success. The study materials that will be given to a candidate will be accompanied by a description of what is considered to be basic skills and knowledge that is required. The local Facilities Supervisor is responsible for the initial evaluation. Further evaluation will be performed by the incumbent Certified Duty Engineers as part of this program. Their findings will be documented and communicated to local Facilities Supervisor.

Minimum Acceptable Knowledge:
To achieve success, the candidate needs to understand the minimum knowledge that is required to safely operate and maintain the equipment and systems that support the data center and the entire facility. Note that this says the *minimum knowledge required*. Knowing or being able to do more is encouraged, but a candidate must learn what is needed to keep their installation running and what has to be done if something goes wrong. The objective is to minimize, or eliminate, downtime that negatively impacts XXXXX's ability to conduct business.

Training Plan and Schedule:
A training plan shall be prepared by the local Facilities Supervisor and given to the candidate after the initial, and subsequent, candidate evaluations. The plan will outline the flow path from start to final certification for the position that is being pursued by the new candidate. This plan will include the basic knowledge and performance levels required for the position in question. It will outline the expected progress on a weekly basis, identify the name of a previously certified engineer he/she is responsible to work with and tracking the candidate's performance and progress. It will also define the amount of time the candidate is authorized to spend on the program during the working hours each week. The amount of time allocated to training depends upon the candidate's abilities and the judgment of the local Facilities Supervisor. Initially, it is estimated that the new candidate would work with a Certified Duty Engineer in a training atmosphere, 4 to 6 hours a day, for the first 10 to 12 weeks.

Training Materials and Methods:
Employees will be provided study and reference materials, to aid in their certification. These will include basic drawings of the building architecture, electrical and mechanical layouts and basic schematics, with access to AutoCAD drawings, for in-depth study and research.

A candidate for an engineering position will be assigned to an engineer already certified, as previously stated. The candidate will work with this Certified Duty Engineer for a four-week period, or a period set by the local Facilities Supervisor. At the completion of this period, the candidate will be assigned to another engineer for another four weeks. This pattern will be repeated, until the candidate has had the opportunity to work with each engineer in that group. By following this pattern, the candidate will be able to see all aspects of the position being sought and will be able to learn from each engineer. At the end of each period, the assigned Certified Duty Engineer and local Facilities Supervisor will evaluate the candidate's performance and abilities. Together, they will highlight strengths and weaknesses and explain to the candidate areas where improvement is needed to bring the candidate's abilities up to the minimum level required for placement on the duty engineer rotation.

Evaluation of Progress:

Evaluation of the candidate's progress is an essential part of the certification process, to determine the level that the candidate has reached and to aid them in attaining certification. This should occur at approximately 2-week intervals, and will consist of whatever the Certified Duty Engineer that is currently working with the candidate deems necessary. This can be any form of testing that is deemed effective; true/false, multiple choice, essay, drawings with blank sections to be filled in, practical hands-on testing, some combination of these, or an alternative approved by the Local Facilities Supervisor.

Evaluation can be expected to occur throughout the certification process, and may cover all topics that the candidate has been exposed to up to that date. For example, at week six, all the information from the previous six weeks may be covered, not just the information learned during the last evaluation period. Part of the evaluation process may also include a series of set-points and parameters, such as voltages, room names, equipment nomenclatures, capacities, sizes, etc.

Feedback from the Local Facilities Supervisor is important to the candidate, and to the group, to ensure that a solid base is established within the group for the candidate and/or, to establish the candidate's suitability for the job. A poor match between the job and the candidate will only hurt the group as a whole, may jeopardize date center operations and leave a candidate frustrated and in a position to fail. *Therefore, an honest evaluation by the Certified Duty Engineers and the Local Facilities Supervisor must be a priority. The candidate needs to be informed as to their strengths, as well as their weaknesses, so that they can be shown their progress and areas that may require improvement.*

If a candidate is determined to be unable to grasp the required knowledge or develop the necessary skills necessary for them to support the Duty Engineer Rotation, then the Facilities Department will not retain the candidate. The decision to release a candidate from employment may be based upon evaluations done at the fourth week, eighth week, or at the completion of the twelfth week, but all possible opportunities to finish the certification process will be made available for new hires prior to, and up to the end of this probationary period.

At the completion of twelve weeks instruction and on-the-job training, the candidate then has an additional period of time, up to twenty-four weeks, to complete **all** requirements to be fully supporting the Duty Engineer Rotation. If, at the end of the <u>MAXIMUM</u> twenty-four week period, the candidate is not ready to be placed on the Duty Engineer Rotation, the candidate will be evaluated again for either release from employment, or for additional training. The Management will make this decision.

Qualification for Advancement:

The candidate must also understand that completion of the certification for the Duty Engineer Rotation, is not a guarantee for advancement or placement beyond the initial hiring position. Completion of the training and a candidate's performance will, however, be taken into consideration for following assignments and merit programs.

Work Assignments and Training Time:

During the course of the candidate's training, time will be made available for self-study, field assignments, facility projects, and working with the facility preventative maintenance program. Field assignments, projects, and PM's should all be designed to aid candidates in finding their way around the facility, locating equipment, learning how to performance maintenance, and respond to emergency situations. In addition, the time should be used to developing an understanding of the facility's mission, and how their job performance is critical in the support of the data center availability. This shall be a coordinated plan between the assigned Certified Duty Engineer and the Local Facilities Supervisor.

All time spent during the pursuit of Duty Engineer Certification will be documented by will be documented via CATS (or another appropriate time-management program).

Completion of Certification and Re-certification:

Once a candidate is certified, the certification process does not stop. Every certified candidate will be required to re-certify every two years. This recertification is designed to help each individual remain current with changes in equipment, policies and procedures, and promote an atmosphere of continuous improvement.

The re-certification progress will be outlined by the Local Facilities Supervisor and will be managed in the same manner as initial certification. Progress reviews for re-certification will be kept at the Local Facilities Supervisor's discretion, based upon the timetable that is established for each individual. The maximum time that will be allowed for an individual to re-certify is eight weeks. If re-certification is not met within this specified time, the Data Center Management will counsel the individual, and corrective measures will be established for the employee to complete the re-certification. As with the initial certification, re-certification will be documented via CATS (or another appropriate time-management program). Items that are mandatory for re-certification are annotated on the SCDP Certification forms by an asterisk (*). Additional items for re-certification may be added by the Local Facilities Supervisor based upon the level of knowledge and skill demonstrated by the candidate. The Local Facilities Supervisor may base this decision upon Supervisor evaluations of the candidate and upon the input of the Facilities staff. The Local Facilities Supervisor may also designate a currently Certified Duty aid in the re-certification process for the candidate, and assist with any corrective measures that may be needed during the candidate's pursuit of recertification.

It is important to note that the SCDP plan is not intended for use as a punitive action against anyone. The Facilities Supervisor will be accountable to oversee this program for the facility they are responsible for, and to ensure that it is not used as a vehicle for harassment or hazing of any kind. The SCDP is not meant to be utilized to isolate or punish any member of the facilities team during initial certification or the two-year follow-up re-certifications. It falls to the Facilities Supervisor, to ensure that this program is used as it is intended, as a training and certification tool, and not as a means to release someone from the facilities staff. Any indication of inappropriate application of this program by anyone must be reported to the Facilities Supervisor or Data Center Operations Manager for resolution.

BASIC FLOW-PATH FOR CERTIFICATION
WEEKS 1 & 2

- Process paperwork for security, introduction to software platforms used in normal activities.
- Tour of facilities
- Code of conduct and mission statement training
- Radio Training (codes used and meanings)
- Training on raised floor requirements and operations
- Familiarization with facility rooms, switchgear identifications/locations, critical equipment identification and operations
- Learn department forms and procedures (VinNet, WorkplaceOnsite, CATS, SDMS, ServiceCenter Client, etc.)
- Review all facilities records (calibration, PM testing, PM scheduling, training, fire inspections, etc.)
- Lockout/Tagout Training
- Office/lifting/general safety training
- Learn facilities logs, equipment histories, time sheets
- Familiarization with BAS and associated systems
- Simple PMs that require no supervision (designed to force candidate to travel installation and find equipment/locations)
- Walkthroughs with assigned engineer on equipment

WEEKS 3 & 4

- Training on electrical systems; including operating procedures and one-line drawings, 'hand-over-hand' tracing of systems from start to end of systems. This includes self-study time.
  - o Utility supply systems
  - o 4160 VAC systems
  - o 480 VAC systems
  - o 120/208 VAC systems
  - o Emergency Generator systems
  - o Battery systems
- Training on HVAC and associated systems; including operating procedures and one-line diagrams, 'hand-over-hand' tracing of the systems from start to end of systems. This includes self-study time.
  - o CRAC systems
  - o Heat exhaust fans
  - o Air-tiles
- Training on the facilities layout, electrical systems, security

systems and plumbing systems

• Training of equipment movement; pallet jack, power lift, dock plates

• Fire extinguisher training and fire-suppression systems training

• PM's that require little or no supervision; more demanding PM's may be done with the assigned engineer or any other engineer of that group, if approved by the Facilities Supervisor. These PM items should be designed to give the candidate more responsibility and give greater opportunity for familiarization with the facility.

• Small project work that requires the candidate to pursue planning, ordering parts and/or service, supervision of contractors, and some work requiring a demonstration of work skills

• 30-day evaluation of candidate with Data Center Facilities Management

WEEKS 5 through 8

• Electrical Safety Training

• Assigned areas of responsibility for PM's, maintenance, upkeep, inventories control. These will be in line with what the candidate's hired position is and will be assigned by the Local Facilities Supervisor

• Training on electrical components operation, construction, and specific PM's associated with the electrical equipment; this includes self-study time.

• Training on the UPS systems and equipment operations, standard and backup operating procedures, back-ups, specific PM's and battery interfaces, static auto-tie purpose and operations; this includes self-study time.

• Training and orientation for in-depth operation and procedures of the BAS system, assignment of passwords for limited access to the BAS system and its host programming

• Training on procedures to power up and power down equipment with an emphasis on safety and continued site operations; this includes self-study time.

• Training on the security systems and procedures

• Walk-throughs with an assigned engineer on equipment and procedures

• Training on the SPCC Plan (Spill Prevention Control and Countermeasures Plan)

WEEKS 9 through 12

- Water distribution training, options and emergency actions; this includes self-study time.
- Assignment of smaller projects to become familiar with site procedures, interaction with vendors, documentation requirements, and to allow development of skills needed to successfully complete a project.
- Hands-on fire training
- Training and orientation for in-depth preventative maintenance on the switchgear and UPS equipment.
- Training procedures to power up and to power down equipment with an emphasis on safety and continued site operations; this includes self-study time.
- Training on stored-energy and associated systems.
- Training on mechanical equipment safety and operations.
- At the Local Facilities Supervisor's discretion, the trainee may begin "under instruction" training as the Duty Engineer, by being assigned this duty with the qualified engineer.

WEEKS 13 through 18

During this time, the individual will be a functioning member of their facilities group and be assigned work and responsibilities in accordance with their abilities and demonstrated skills. They will continue to be assigned to a qualified engineer to work with but will be expected to perform on their own. Additionally, they will be assigned to the Duty Engineer rotation, with a qualified engineer, to become experienced with the routines of the position as well as being exposed to the responsibilities for that position.

At the end of this period, ideally, the individual will be ready to be place into the Duty Engineer position as a qualified Duty Engineer. This will be at the discretion and better judgment of the engineering group, with final approval by the Data Center Facilities Management. IF the individual is not ready at this time, additional time may be allotted to them, as approved by the Data Center Facilities Management, up to the end of the six months probationary period. At that time, if an individual cannot be qualified for Duty Engineer, actions shall be taken by the Data Center Facilities Management to either release the individual, or transfer them into a department where they may be better suited.

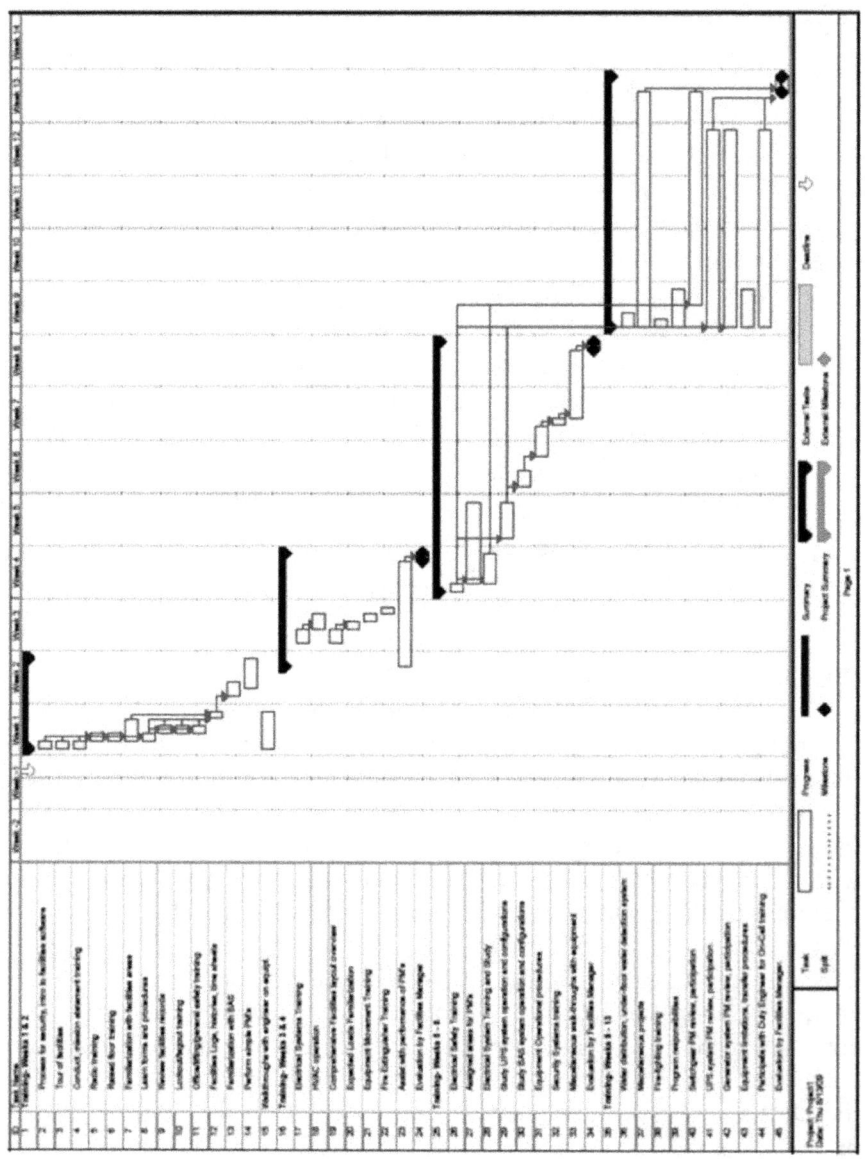

SCDP CERTIFICATION FOR DUTY ENGINEER
APPENDIX A
• XXXXX DATA CENTER SCDP LEVEL CRITERIA

A: ENTRY LEVEL – Duty Engineer Candidate
- Possess basic electrical skills and demonstrate the ability to assist with power cable installation
- Possess basic mechanical skills and demonstrate the ability to perform basic mechanical equipment PMs.
- Demonstrate working knowledge in the use of test equipment
- Demonstrate working knowledge of power and hand tool use and safety
- Demonstrate basic PC skills – word, excel, windows
- Demonstrate sound knowledge of the local MSDS program
- Demonstrate sound knowledge of lockout/tagout procedures
- At least two years' experience in Electrical or HVAC
- Familiar with raised floor operation
- Demonstrate basic CATS system skills – enter time, and create work requests
- Possess basic administrative/project management skills and demonstrate the ability to organize and plan tasks as assigned by the Facilities Supervisor.

B: MATURE LEVEL – Duty Engineer
- Read and understand all drawings, electrical schematics, piping diagrams, structural design plans, etc.
- Operational knowledge of the change management system
- Operational knowledge of the SCS ticket management system
- Operational knowledge of Asset-Center
- Proficient in the use of mechanical and test equipment
- Operational knowledge of electrical one-line distribution for the site
- Installation/Removal of power cables for computer room equipment at the direction of the Facilities Supervisor
- Operational knowledge of the fire monitoring and suppression systems, to include FM-200 and sprinklers
- Operational knowledge of CRAC unit operation
- Operational knowledge of building water supplies, and shutoff

217

locations

• Operational knowledge of all systems used for remote monitoring of data center facilities

• Ability to supervise contracted mechanical and electrical projects as assigned by the local Facilities Supervisor

• Operational knowledge of physical building security

• Operational knowledge of building pneumatic systems

• Proficient in all aspects of building emergency operation

• Perform on-call facilities duties

• Certified CFC handling in accordance with Federal Regulation 10 CFR 81, Subpart F

• Understands, administers and controls contractor and field services for work in and on data center equipment

• Demonstrates a sound knowledge of XXXXX and Asset Management practices and procedures, as they apply to facilities maintenance

C: ADVANCED LEVEL – Senior Engineer

• Demonstrate sound knowledge of all facilities systems

• Isolate, troubleshoot and perform lead duties during repair of generators, CRAC units, switchgear and UPS systems

• Perform lead duties during major PM work

• Insure compliance to XXXXX/OSHA guidelines

• Perform manpower planning for all major PMs

• Perform load trend analysis for all major distribution systems

• Insure all power requirements are installed and recorded on tracking systems

• Develop and maintain AutoCAD as-built drawings and schematics for all support equipment

D: EXPERT LEVEL – Senior Engineer

• Possess sound knowledge of entire building operations and equipment

• Assist monitoring budget for data center facilities, per direction of the Facilities Supervisor

• Administer purchasing/invoicing systems, per direction of the Facilities Supervisor

• Manage employees in department, per direction of the Facilities Supervisor

- Ensure adherence to all XXXXX practices and OSHA guidelines pertaining to safety and operations
- Establish and maintain good working relations with data center management
- Ensure training goals are met, per direction of the Facilities Supervisor
- Evaluate sufficient infrastructure exists for future growth/projects, per direction of the Facilities Supervisor
- Assist the Facilities Supervisor to produce monthly operations reports and verify accuracy of results

NOTE:

ENTRY LEVEL- performs duties with supervision
MATURE LEVEL- performs duties with minimal supervision, supports duty rotation
ADVANCED LEVEL- performs and teaches with little or no supervision
EXPERT LEVEL- performs and teaches without supervision

Facilities **Position Skills Inventory** General **Knowledge/Administration**	Skill Level	Skill Level	Skill Level	Skill Level
Functional Skills Pertaining to: Facilities	Entry	Mature	Advanced	Expert
Facilities Functional Competencies				
General				
ServiceCenter Client, Windows XP/Office	I	I	I	A
Basic Troubleshooting and Repair	A	A	A	A
Blueprint Reading and Updates (Building)	B	I	A	A
Change management process	B	I	I	A
Ticket management process	B	I	I	A
Computer Room General Work and Safety Procedures	B	I	A	A
Contractor Management	B	I	I	A
- Janitorial				
- Security				
- BTS				
Remote Monitoring of Facilities	B	I	I	A
Procurement Process	B	I	I	A
Physical Security				
Administration of Ingress	B	I	I	A
Physical Security Policies and Procedures	B	I	I	A
Troubleshooting and Repair of Controls	I	I	I	A

Troubleshooting and Repair of Door Hardware	I	A	A	A
Fire Alarm and Suppression Systems				
Operation	B	I	A	A
Maintenance	B	I	I	A
Troubleshooting and Repair	B	I	I	A
Safety				
MSDS Program	I	A	A	A
Lockout/Tagout	I	A	A	A
Electrical Safety	B	A	A	A
Mechanical Safety	B	A	A	A
High Energy System Safety	B	A	A	A
CPR/AED Training	B	A	A	A

B = Basic (able to perform task with supervision)
I = Intermediate (able to perform task with minimal supervision and/or engineering or vendor assistance)
A = Advanced (able to perform and/or teach task with no supervision

Facilities Position Skills Inventory Mechanical Skills	Skill Level	Skill Level	Skill Level	Skill Level
Functional Skills Pertaining to: **Facilities**	Entry	Mature	Advanced	Expert
Facilities Functional Competencies				
Mechanical				
Basic Troubleshoot and Repair	I	A	A	A

Capacity Planning	B	I	A	A
Code Application	B	I	A	A
Distribution	B	I	A	A
Indoor Air Quality Administration	B	I	A	A
Load Management/Trend Analysis	B	I	A	A
Project Management	B	I	A	A
Theoretical Application - Mechanical	B	I	A	A
Use of Tools and Test Equipment	I	A	A	A
HVAC System				
Operation	B	I	A	A
Maintenance	B	I	I	A
Troubleshooting and Repair	B	I	I	A
CFC Certification	B	A	A	A
Building Automation System and Controls				
Operation	B	I	A	A
Maintenance	B	I	I	A
Troubleshooting and Repair	B	I	I	A
Pneumatics				
Maintenance	I	A	A	A
Operation	I	A	A	A
Repair	I	A	A	A
Mechanical Safety	I	A	A	A
High Energy System Safety	I	A	A	A
B = Basic (able to perform task with supervision)				
I = Intermediate (able to perform task with minimal supervision				

and/or engineering or vendor assistance) A = Advanced (able to perform and/or teach task with no supervision				

Facilities Position Skills Inventory Electrical Skills	Skill Level	Skill Level	Skill Level	Skill Level
Functional Skills Pertaining to: Facilities	Entry	Mature	Advanced	Expert
Facilities Functional Competencies				
Electrical				
Basic Troubleshoot and Repair (Onelines)	I	A	A	A
Capacity Planning	B	I	A	A
Electrical Code Application	I	I	A	A
Distribution	B	I	A	A
Power Cable Management	B	I	A	A
Load Management/Trend Analysis	B	I	A	A
Project Management	B	I	A	A
Theoretical Application - Electrical	B	I	A	A
Use of Tools and Test Equipment	I	A	A	A
UPS System				
Operation	B	I	A	A
Maintenance	B	I	I	A
Troubleshooting and Repair	B	I	I	A

Static Auto-Tie System				
Operation	B	I	A	A
Maintenance	B	B	I	A
Troubleshooting and Repair	B	B	I	A
Generator System				
Operation	B	I	A	A
Maintenance	B	I	I	A
Repair	B	I	I	A
Switchgear				
Operation	B	I	A	A
Maintenance	B	I	I	A
Repair	B	I	I	A
B = Basic (able to perform task with supervision) I = Intermediate (able to perform task with minimal supervision and/or engineering or vendor assistance) A = Advanced (able to perform and/or teach task with no supervision				

SCDP CERTIFICATION FOR DUTY ENGINEER
APPENDIX B
XXXXX DATA CENTER SCDP LEVEL CRITERIA
Section 1: Reading Assignments

For an Engineer candidate to be properly prepared for being ready to support the Duty Engineer rotation, the candidate needs to develop a certain level of knowledge. Part of this knowledge can only be found in several documents that are XXXXX documents, or under the cognizance of some outside agency, but are still required knowledge. This type of knowledge can only be obtained by the candidate reading the material and developing a sound understanding.

Each item listed below will be read by the candidate and then signed as being understood by the candidate.

> I.XXXXX Lockout/Tagout procedures (Compliance Manual, The Control of Hazardous Energy 29 CFR 1910.147)

_____
*

Candidate Signature   Date

> II.XXXXX Occupational Safety & Health Policies

_____

Candidate Signature   Date

> III.Employee Safety and Industrial Hygiene Standard

_____

Candidate Signature   Date

XXXXX BCBS IT Computing Facility Recovery Manual Restoration and Salvage Team Plan

_____

Candidate Signature   Date

XXXXX Hazard Communication Plan

_____
*

Candidate Signature   Date

> IV.XXXXX Peregrine Change Management Process

_____
*

Candidate Signature   Date

> V.XXXXX SDMS Procedures

_____
*

Candidate Signature   Date

> VI.XXXXX Service Center Client 6.1.5 (Ticket Management) Process

_____
*

Candidate Signature   Date

DC-SOP-05 BBS NA ITO Data Center Electrical Safety Policy and Procedures
(Compliance Manual for OSHA Standard 29 CFR 1910.331-.335)

_____*
Candidate Signature   Date

VII. State Code of Regulations as applicable to Facility Operations
(Local state codes for area facility is in)

_____
Candidate Signature   Date

VIII.XXXXX Data Center Maintenance and Work Rules

_____
Candidate Signature   Date

IX.DC-SOP-06 BBS NA ITO Standard Operating Procedure for Data Center Equipment Grounding - Pittsburgh

_____*
Candidate Signature   Date

X.MAINTENANCE PROCEDURE "Raised Floor Area Electrical Cable Installations" Procedure No. MEU 11-7

_____
Candidate Signature   Date

XI.DC-SOP-04 BBS NA ITO Data Center Procedure for Verification and Installation of Power Supplies to Non-Rack Mount Production Equipment

_____
Candidate Signature   Date

XII.SCM-SOP-10: North America IT Standard Operating Procedure for Service Continuity Management Team Audit of Recovery Plans

_____
Candidate Signature   Date

XXXXX DATA CENTER SCDP LEVEL CRITERIA

Section 2: General Policies and Training

The following sections of the SCDP Certification for Duty Engineer consist of performance factors, demonstrations of knowledge and skills, and follow-up documentation by designated Certified Personnel. For each item listed, they must either be Performed (P), Simulated/Demonstrated (S), Discussed (D), or Observed (O). Several items may have more than one designated way of accomplishment, and the priority for these is in the following order: P, S, D, O.

A. General Knowledge/Practical Factors:

1. Able to use MS-Office/Windows _____ (P,S)
Certified Duty Engineer/Date

2. Basic Mechanical Troubleshooting Techniques and Procedures
_____ (P,S,D)
Certified Duty Engineer/Date

4. Basic Electrical Troubleshooting Techniques and Procedures
_____ (P,S,D)
Certified Duty Engineer/Date

5. Ability to read and interpret building blueprints
_____ (P,S)
Certified Duty Engineer/Date

5. Ability to read and interpret electrical schematics and drawings
_____ (P,S)
Certified Duty Engineer/Date

6. Ability to read and interpret mechanical piping diagrams and drawings
_____ (P,S)
Certified Duty Engineer/Date

7. Aperture system operation, documentation and record keeping.
_____ (P)
Certified Duty Engineer/Date

8.   Work Requests and procurement procedures for vendors and BTS
   _____(P,S)
Certified Duty Engineer/Date

9.   Vendor Management for work in the Data   Center Facilities
   _____(P,S)
Certified Duty Engineer/Date

SCDP CERTIFICATION FOR DUTY ENGINEER

9.   Vendor Management for work in the Data   Center Facilities
_____(P,S,D)
Certified Duty Engineer/Date

10.   Procedures for raised floor work in the computer rooms
_____(P,S,D)
Certified Duty Engineer/Date

11.   Change management and control
_____(S,D,O)
Certified Duty Engineer/Date

12.   Physical Security Procedures and policies for the facility
_____(D)
Certified Duty Engineer/Date

13.   Operation and maintenance for the Fire Alarm Systems
_____(D)
Certified Duty Engineer/Date

14.   Operation and maintenance for the Fire
Suppression System   _____(D)
Certified Duty Engineer/Date

15.   Understanding of the MSDS System   _____(D)
Certified Duty Engineer/Date

16.   Understanding of the Hazardous   Communications Program
_____(D)
Certified Duty Engineer/Date

17.   Understanding of the Facilities Lockout/Tagout Procedures _-
_____(P,S,D)
Certified Duty Engineer/Date

18.   Understanding of Electrical Safety Procedures and Principles
_____(P,S,D)
Certified Duty Engineer/Date

19. Understanding of Mechanical Safety Procedures and Principles

_____(P,S,D)
Certified Duty Engineer/Date

20. Concerns and precautions associated with High Energy Systems

_____(P,S,D)
Certified Duty Engineer/Date

SCDP CERTIFICATION FOR DUTY ENGINEER

21. CPR/AED Certification   _-
_____Certified Duty Engineer/Date.

B1. Basic Electrical Knowledge/Practical Factors:

1. In-depth Electrical Safety knowledge as it pertains to Data Center equipment and installations
   _____(P&D)
   Certified Duty Engineer/Date

2a. AC Electrical Theory, as it pertains to Data Center equipment and installations
   _____(D)
   Certified Duty Engineer/Date

2b. DC Electrical Theory, as it pertains to Data Center equipment and installations
   _____(D)
   Certified Duty Engineer/Date

3. Detailed electrical troubleshooting procedures
   _____(P,S,D)
   Certified Duty Engineer/Date

4. Understanding of the Utility-supplied Electrical Distribution Systems
   _____*(P,S,D)
   Certified Duty Engineer/Date

5. Understanding of the 480VAC Electrical Distribution Systems
   _____*(P,S,D)
   Certified Duty Engineer/Date

6. Understanding of the 277VAC Electrical Distribution Systems
   _____*(P,S,D)
   Certified Duty Engineer/Date

7. Understanding of the 208/120VAC Electrical Distribution Systems
   _____*(P,S,D)
   Certified Duty Engineer/Date

8. Understanding of the DC Electrical Distribution Systems
   _____(P,S,D)
   Certified Duty Engineer/Date

9. Describe the principles of load management for the Data
   _____*(D)
   Certified Duty Engineer/Date

Center Facility

10. Demonstrate in-depth knowledge of the National Electric Code (NEC), as it pertains to the Data Center Facility

_____*(D)
Certified Duty Engineer/Date

11. Demonstrate a working knowledge of all electrical test equipment used at the Facility

_____*(P,S,D)
Certified Duty Engineer/Date

SCDP CERTIFICATION FOR DUTY ENGINEER

B2. UPS Knowledge/Practical Factors:

1. In-depth operation of the UPS modules
    _____(P&D)
    Certified Duty Engineer/Date

2. Understanding of the UPS Electrical Distribution Systems
    _____(P&D)
    Certified Duty Engineer/Date

3. Secure UPS Module
    _____(P)
    Certified Duty Engineer/Date

4. Start UPS Module
    _____(P)
    Certified Duty Engineer/Date

5. Assist and perform Preventative Maintenance on UPS Modules
    _____(O)
    Certified Duty Engineer/Date

    _____(O)
    Certified Duty Engineer/Date

    _____(P)
    Certified Duty Engineer/Date

    _____(P)
    Certified Duty Engineer/Date

6. Shift UPS Loads from dual-loop to Single Loop and back, via Static Auto-Tie
    _____(P,S,D)
    Certified Duty Engineer/Date

7. Understanding of the UPS Battery Systems
    _____*(D)
    Certified Duty Engineer/Date

8. Demonstrate ability to charge a UPS Battery
    _____(P,S,D)
    Certified Duty Engineer/Date

SCDP CERTIFICATION FOR DUTY ENGINEER

B3. Emergency Generator (EG) Knowledge/Practical Factors:

1. In-depth operation of the Emergency Generators

    _____*(P&D)
    Certified Duty Engineer/Date

2. Understanding of the Emergency Generator Electrical Distribution Systems

    _____*(D)
    Certified Duty Engineer/Date

3. Assist in the performance of Preventative Maintenance on Emergency Generators

    _____(O)
    Certified Duty Engineer/Date

    _____(O)
    Certified Duty Engineer/Date

    _____(P,O)
    Certified Duty Engineer/Date

    _____(P,O)
    Certified Duty Engineer/Date

5. Assist and perform weekly/monthly operational testing of the Emergency Generators

    _____(O)
    Certified Duty Engineer/Date

    _____(O)
    Certified Duty Engineer/Date

    _____(P)
    Certified Duty Engineer/Date

    _____(P)
    Certified Duty Engineer/Date

6. Understanding of the Emergency Generator Battery Systems

    _____*(D)
    Certified Duty Engineer/Date

7. Understanding of the UPS Battery Systems

    _____*(D)
    Certified Duty Engineer/Date

8. Demonstrate ability to charge an Emergency Generator Battery

_____(P,S,D)
Certified Duty Engineer/Date

SCDP CERTIFICATION FOR DUTY ENGINEER

B4. Electrical Switchgear Knowledge/Practical Factors:

1. In-depth operation and construction of 480VAC switchboards

    _____*(P&D)
    Certified Duty Engineer/Date

2. Cross-connect Load Center Switchboards

    _____(P,S)
    Certified Duty Engineer/Date

3. Assist and perform Preventative Maintenance on Distribution Load Centers

    _____(O)
    Certified Duty Engineer/Date

    _____(O)
    Certified Duty Engineer/Date

    _____(P)
    Certified Duty Engineer/Date

    _____(P)
    Certified Duty Engineer/Date

4. Understanding of the Load Center Control Power System

    _____*(P,S,D)
    Certified Duty Engineer/Date

5. Demonstrate an in-depth knowledge of Load Center Breaker operation and construction

    _____(P,S,D)
    Certified Duty Engineer/Date

6. Demonstrate ability and knowledge to troubleshoot Load Center Breakers

    _____(P,S,D)
    Certified Duty Engineer/Date

SCDP CERTIFICATION FOR DUTY ENGINEER

B5. Automatic Transfer Switch (ATS) Knowledge/Practical Factors:

1. In-depth operation of the Automatic Transfer Switches

    _____*(P&D)
    Certified Duty Engineer/Date

2. Understanding of the Electrical Distribution Systems supplied by the Automatic Transfer Switches

    _____*(D)
    Certified Duty Engineer/Date

3. Assist in the performance of Preventative Maintenance on the Automatic Transfer Switches

    _____(O)
    Certified Duty Engineer/Date

    _____(O)
    Certified Duty Engineer/Date

    _____(P,O)
    Certified Duty Engineer/Date

    _____(P,O)
    Certified Duty Engineer/Date

5. Assist and perform weekly/monthly operational testing of the Automatic Transfer Switches

    _____(O)
    Certified Duty Engineer/Date

    _____(O)
    Certified Duty Engineer/Date

    _____(P)
    Certified Duty Engineer/Date

    _____(P)
    Certified Duty Engineer/Date

SCDP CERTIFICATION FOR DUTY ENGINEER

C1. Basic Mechanical & HVAC Knowledge/Practical Factors:

1. In-depth Mechanical Safety knowledge as it pertains to Data Center equipment and installations

    _____(P&D)
    Certified Duty Engineer/Date

2. HVAC Theory, refrigeration cycle and how it pertains to Data Center equipment

    _____(D)
    Certified Duty Engineer/Date

3. Detailed mechanical troubleshooting procedures

    _____(P,S,D)
    Certified Duty Engineer/Date

4. Basic understanding of the Computer Room Air Conditioner (CRAC) units

    _____(P,S,D)
    Certified Duty Engineer/Date

5. Basic understanding of the Liebert Site-Scan system, how it functions and monitors functions

    _____(P,S,D)
    Certified Duty Engineer/Date

6. Understanding of the Ventilation system lay-out and functions

    _____(P,S,D)
    Certified Duty Engineer/Date

7. Demonstrate a basic knowledge of the applicable sections of the OSHA standards and NFPA Codes

    _____(P,S,D)
    Certified Duty Engineer/Date

8. Demonstrate an understanding of tools and their uses

    _____(P,S,D)
    Certified Duty Engineer/Date

9. Demonstrate in-depth knowledge of the HVAC cooling strategies, and how they pertain to Data Center equipment and installations

    _____*(D)
    Certified Duty Engineer/Date

10. Shift heat loads from one CRAC to another

    _____*(P&D)
    Certified Duty Engineer/Date

11. Assist and perform

    _____(O)

Preventative Maintenance on
CRAC units

Certified Duty Engineer/Date

_____(O)
Certified Duty Engineer/Date

_____(P,O)
Certified Duty Engineer/Date

_____(P,O)
Certified Duty Engineer/Date

SCDP CERTIFICATION FOR DUTY ENGINEER

D1. Abnormal/Casualty Operations Knowledge/Practical Factors:

This section of the certification process addresses the most probable types of failures and casualties that the candidate is most likely to be faced with. It does NOT, by any means, outline all the possible scenarios that may occur at the facility. There are any number of possible situations that may arise, not included in this general listing, which will require the candidate to be to act in a crisis, analyze the situation facing them, and to take appropriate actions to maintain the data center at 100% capability.

Demonstrate in-depth knowledge of immediate actions to be taken for each of the following situations:

1. Loss of utility electric power
    to the site

    _____*(P,S,D)
    Certified Duty Engineer/Date

2. Loss of CRAC unit

    _____*(P,S,D)
    Certified Duty Engineer/Date

3. Loss of UPS

    _____*(P,S,D)
    Certified Duty Engineer/Date

4. Fire in the Data Center

    _____*(P,S,D)
    Certified Duty Engineer/Date

5. Extreme inclement weather

    _____*(P,S,D)
    Certified Duty Engineer/Date

6. Loss of PDU   _____*(P,S,D)
           Certified Duty Engineer/Date

7. Loss of Rack Power Strip   _____*(P,S,D)
           Certified Duty Engineer/Date

8. Personnel Electrocution   _____*(P,S,D)
           Certified Duty Engineer/Date

9. Diesel Fuel Spill   _____*(P,S,D)
           Certified Duty Engineer/Date

10. Battery Acid Spill   _____*(P,S,D)
           Certified Duty Engineer/Date

11. Battery Thermal Runaway   _____*(P,S,D)
           Certified Duty Engineer/Date

SCDP CERTIFICATION FOR DUTY ENGINEER

D1. Abnormal/Casualty Operations Knowledge/Practical Factors:

12. Failure of Utility Main   _____*(P,S,D)
     Breakers   Certified Duty Engineer/Date

13. Failure of 480 VAC Breakers   _____(P,S,D)
           Certified Duty Engineer/Date

14. Loss of Switchgear Control   _____(P,S,D)
     Power   Certified Duty Engineer/Date

15. Failure of Switchgear   _____(P,S,D)
     Automatic Functions   Certified Duty Engineer/Date

16. Failure of 4160/480 VAC   _____(P,S,D)
     Transformer   Certified Duty Engineer/Date

17. Failure of 480/120 VAC   _____(P,S,D)
     Transformer   Certified Duty Engineer/Date

18. Failure of Emergency Generator Main Breaker During Emergency Generator Start/loss of power

_____(P,S,D)
Certified Duty Engineer/Date

19. Failure of Automatic Transfer Switch During loss of power

_____(P,S,D)
Certified Duty Engineer/Date

20 Loss of one side of the electrical power distribution system

_____(P,S,D)
Certified Duty Engineer/Date

SCDP CERTIFICATION FOR DUTY ENGINEER

E1. Certification Complete

The following list of courses and schools are recommended for certification. These are not mandatory and may be marked as not applicable by the Facilities Supervisor, based upon the candidate's experience, demonstrated skills, and facilities staff input.

1. BAS

   _____
   Certified Duty Engineer/Date

2. HVAC/Refrigeration

   _____
   Certified Duty Engineer/Date

3. CFC Certification

   _____
   Certified Duty Engineer/Date

4. CPR/AED Certification

   _____
   Certified Duty Engineer/Date

5. Motor and Motor Controls

   _____
   Certified Duty Engineer/Date

6. Electrical Troubleshooting

   _____
   Certified Duty Engineer/Date

7. Electrical Switchgear/Distribution

   _____
   Certified Duty Engineer/Date

8. UPS

   _____
   Certified Duty Engineer/Date

9. Electrical Safety

   _____
   Certified Duty Engineer/Date

10. Auto Static-Tie

   _____
   Certified Duty Engineer/Date

SCDP CERTIFICATION FOR DUTY ENGINEER

F1. Recommended Courses/Schools:

The following signatures annotate areas of the Duty Engineer Certification that have been accomplished. Upon completion of all signatures, and final signature of the Facilities Supervisor, this page will be in the candidate's personnel record as proof of certification.

Section 1:	Reading Assignments	_____ Certified Duty Engineer/Date
Section 2:	General Policies and Training	
B1.	Basic Electrical Knowledge/Practical Factors	_____ Certified Duty Engineer/Date
B2.	UPS Module Knowledge/Practical Factors	_____ Certified Duty Engineer/Date
B3.	Emergency Generator Knowledge/Practical Factors	_____ Certified Duty Engineer/Date
B4.	Electrical Switchgear Knowledge/Practical Factors	_____ Certified Duty Engineer/Date
B5.	Automatic Transfer Switch (ATS) Knowledge/Practical Factors:	_____ Certified Duty Engineer/Date
C1.	Basic Mechanical Knowledge/Practical Factors	_____ Certified Duty Engineer/Date
C2.	HVAC Knowledge/Practical Factors	_____ Certified Duty Engineer/Date
D1.	Abnormal/Casualty Operations Knowledge/Practical Factors	_____ Certified Duty Engineer/Date

E1.   Recommended
       Courses/Schools   _____

                                     Certified Duty Engineer/Date

Candidate ready for Duty Engineer Rotation

_____

Facilities Supervisor/Date

APPENDIX C

SCDP CERTIFICATION FOR DUTY ENGINEER

SCDP Certification for Duty Engineer Signature Card Standard

B1. Basic Electrical Knowledge/Practical Factors:

1.	In-depth Electrical Safety knowledge	Discuss PPE requirements for energized work in Class 0, 1 and 2 environments. Describe/discuss test procedures to check "dead" equipment. Demonstrate proper checking of electrical safety gloves, frequency, requirements. Discuss two-man rule. Discuss BBS NA ITO Data Center Electrical Safety Policy and Procedures.
2a.	AC Electrical Theory, as it pertains to Data Center equipment and installations	Demonstrate knowledge of different methods to generate AC power. Discuss generator theory. Demonstrate knowledge of power factor, reactive load and how it pertains to the data center. Demonstrate knowledge of Ohms Law, and how to perform calculations with it for single and 3-phase power. Discuss utility phase variance issues, and how it effects data center operation and reliability.
2b.	DC Electrical Theory, as it pertains to Data Center equipment and	Demonstrate knowledge of different methods to generate DC power. Discuss battery theory, including chemical reactions that create electron flow. Demonstrate knowledge of all DC systems within the Data Center.

installations

3.	Detailed electrical troubleshooting procedures	
4.	Understanding of the Utility-supplied Electrical Distribution Systems	Discuss how utility power is delivered to Data Center. Discuss possible events that might occur, to cause loss of utility power to data center. Discuss how phase variance issues of utility grids affects utility feeds at block-house and data center reliability.
5.	Understanding of the 480VAC Electrical Distribution Systems	Demonstrate in-depth knowledge of one-line diagram of electric plant. Demonstrate in-depth knowledge of switchgear line-ups on most likely abnormal/casualty conditions. Describe how each line-up affects data center reliability.
6.	Understanding of the 277VAC Electrical Distribution Systems	Discuss 277VAC electrical distribution system in the data center.
7.	Understanding of the 208/120VAC Electrical Distribution Systems	Demonstrate in-depth knowledge of 208/120VAC Electrical distribution system. Explain importance of this system to data center mission. Describe fail-over principles, and how it pertains to the data center mission.
8.	Understanding of the DC Electrical Distribution Systems	Demonstrate in-depth knowledge of DC Electrical Distribution System. Discuss procedures to isolate sections of DC electrical system for maintenance or repairs. Discuss specific PPE requirements when working with this system.
9.	Describe the principles of load management	Describe load management. Describe "selective tripping," and how it applies to data center reliability. Describe redundancy measures, nomenclature.

	for the Data Center Facility	
10.	Demonstrate in-depth knowledge of the National Electric Code (NEC) and other associated regulations, as they pertain to the Data Center Facility	Demonstrate in-depth knowledge of NFPA-70, "National Electrical Code® 2008 Edition," NFPA-70E "Standard for Electrical Safety in the Workplace, 2009 Edition," NFPA® 75 "Standard for the Protection of Information Technology Equipment, 2009 Edition"
11.	Demonstrate a working knowledge of all electrical test equipment used at the Facility	

SCDP CERTIFICATION FOR DUTY ENGINEER

SCDP Certification for Duty Engineer Signature Card Standard

B2. UPS Knowledge/Practical Factors:

1.	In-depth operation of the UPS modules	Demonstrate operational knowledge of operations, differences between various types of UPS designs. Demonstrate in-depth knowledge of UPS capacities, alarm meanings and codes, possible problems that may occur and their potential impact on data center reliability. Demonstrate operational knowledge of how to recover from UPS failure, or imminent failure of either a single module, or a critical loop.
2.	Understanding of the UPS Electrical Distribution Systems	Demonstrate operational knowledge of UPS distribution systems. Demonstrate in-depth knowledge of critical bus capacities, alarms, meanings and codes, possible problems that may occur and their potential impact on data center reliability. Demonstrate operational knowledge of auto static-tie operation, discuss potential ramifications on failure of this system, recovery procedures. Demonstrate how phase variance issues effect the operation of the UPS distribution systems, including problems, options to correct these issues and what plant configurations to remedy such issues within the data center. Demonstrate using the one-line diagram
3.	Secure UPS Module	
4.	Start UPS Module	
5.	Assist and perform Preventative Maintenance on UPS Modules	
6.	Shift UPS Loads from dual-loop to	

	Single Loop, via Static Auto-Tie, and back.	
7.	Understanding of the UPS Battery Systems	Demonstrate operational knowledge of how chemical energy is converted to electrical energy, in a lead-acid battery. Demonstrate operational knowledge of physical checks performed to verify battery integrity, including resistance, continuity and specific gravity, and discuss the meaning of results from such testing. Demonstrate operational knowledge of difference between equalize and float charges, and discuss what their purposes are, advantages and disadvantages to battery life and integrity.
8.	Charge a UPS Battery	Perform under the direction of a Powerware technician.

SCDP CERTIFICATION FOR DUTY ENGINEER

SCDP Certification for Duty Engineer Signature Card Standard

B3. Emergency Generator (EG) Knowledge/Practical Factors:

1.	In-depth operation of the Emergency Generators	Demonstrate basic knowledge of how diesel engines work. Demonstrate operational knowledge of different PM's associated with Emergency Generators, and what the results of PM results can mean. Demonstrate operational knowledge of how AC voltage is generated. Demonstrate operational knowledge of Electric Generator PM's and what the expected results of such PM's mean. Demonstrate in-depth knowledge of parameters associated with Emergency Generators (capacities, set-points, start-up times, etc.). Demonstrate operational knowledge of typical maintenance functions, such as weekly startups and what to do if a fuel filter becomes clogged, etc.
2.	Understanding of the Emergency Generator Electrical Distribution Systems	Demonstrate operational knowledge of Emergency Generator power distribution systems. Demonstrate in-depth knowledge of bus capacities, alarms, meanings and codes, possible problems that may occur and their potential impact on data center reliability. Demonstrate operational knowledge of potential ramifications, due to failure of this system and associated recovery procedures.
3.	Assist in the performance of Preventative Maintenance on Emergency Generators	
5.	Assist and perform weekly/monthly operational testing of the Emergency	

Generators

6.	Understanding of the Emergency Generator Battery Systems	Demonstrate operational knowledge of how chemical energy is converted to electrical energy, in a lead-acid battery. Demonstrate operational knowledge of physical checks performed to verify battery integrity, including resistance, continuity and specific gravity, and discuss the meaning of results from such testing. Demonstrate operational knowledge of difference between equalize and float charges, and discuss what their purposes are, advantages and disadvantages to battery life and integrity.
7.	Understanding of the UPS Battery Systems	Demonstrate operational knowledge of how chemical energy is converted to electrical energy, in a lead-acid battery. Demonstrate operational knowledge of physical checks performed to verify battery integrity, including resistance, continuity and specific gravity, and discuss the meaning of results from such testing. Demonstrate operational knowledge of difference between equalize and float charges, and discuss what their purposes are, advantages and disadvantages to battery life and integrity.
8.	Demonstrate ability to charge an Emergency Generator Battery	

SCDP CERTIFICATION FOR DUTY ENGINEER

SCDP Certification for Duty Engineer Signature Card Standard

B4. Electrical Switchgear Knowledge/Practical Factors:

1.	In-depth operation and construction of 480VAC switchgear	Demonstrate operational knowledge of 480 VAC switchboards. Demonstrate how breakers are removed or installed, with emphasis on lifting procedures, safety and PPE. Discuss various switchgear configurations, and demonstrate in-depth knowledge of how each configuration affects data center integrity, reliability and redundancy. Discuss recovery steps, to restore data center integrity when abnormal/casualty event occurs. Heavy emphasis will be place on demonstration to "think on your feet," by using the one-line diagram and also by memory, without the assistance of the one-line diagram.
2.	Cross-connect Load Center Switchboards	
3.	Assist and perform Preventative Maintenance on Distribution Load Centers	
4.	Understanding of the Load Center Control Power System	Demonstrate in-depth knowledge of switchgear and PLC control power system. Discuss ramifications of failure of control power system, recovery steps.
5.	Demonstrate an in-depth knowledge of Load Center Breaker operation	Demonstrate operational knowledge of how breakers work. Demonstrate in-depth knowledge of manual breaker operation, with emphasis on safety procedures, PPE and safety checks to be made before shutting breakers.

	and construction	Demonstrate ability to manually operate breaker. Discuss meaning of different flags on breakers, and ramifications. Demonstrate in-depth knowledge of load capacities, trip settings and why we have different trip settings.
6.	Demonstrate ability and knowledge to troubleshoot Load Center Breakers	

SCDP CERTIFICATION FOR DUTY ENGINEER

SCDP Certification for Duty Engineer Signature Card Standard

B5. Automatic Transfer Switch (ATS) Knowledge/Practical Factors:

1.	In-depth operation of the Automatic Transfer Switches	Demonstrate in-depth knowledge of how ATS systems operate, timing, settings and power ratings. Demonstrate operational knowledge of potential problems that may occur with ATS switches, and immediate actions. Heavy emphasis will be place on demonstration to "think on your feet," by using the one-line diagram and also by memory, without the assistance of the one-line diagram.
2.	Understanding of the Electrical Distribution Systems supplied by the Automatic Transfer Switches	Demonstrate in-depth knowledge all systems supplied by ATS feeds. Describe ramifications of ATS failure, and describe immediate actions, to restore integrity to data center, in the event of an abnormal/casualty event, to any ATS switch. Heavy emphasis will be place on demonstration to "think on your feet," by using the one-line diagram and also by memory, without the assistance of the one-line diagram.
3.	Assist in the performance of Preventative Maintenance on the Automatic Transfer Switches	
5.	Assist and perform weekly/monthly operational testing of the Automatic Transfer Switches	

SCDP CERTIFICATION FOR DUTY ENGINEER

SCDP Certification for Duty Engineer Signature Card Standard

C.   Basic Mechanical & HVAC Knowledge/Practical Factors:

1. In-depth Mechanical Safety knowledge as it pertains to Data Center equipment and installations

    Demonstrate knowledge of mechanically stored energy and safety precautions. Discuss mechanical safety issues that occasionally occur within the data center, appropriate PPE or steps to mitigate such risks, and responsibilities of Facilities Team members to be alert for such risks.

2. HVAC Theory, refrigeration cycle and how it pertains to Data Center equipment

    Demonstrate operational knowledge of the refrigeration cycle. Demonstrate operational knowledge of Latent Heat of Vaporization, and its meaning to the refrigeration cycle. Demonstrate operational knowledge of factors affecting overall heat transfer efficiency in a refrigeration unit, and how we utilize such factors within the data center environment.

3. Detailed mechanical troubleshooting procedures

4. Basic understanding of the Computer Room Air Conditioner (CRAC) units

    Demonstrate operational understanding of DX Air-cooled CRAC units, like those utilized by the XXXXX Data Center. Demonstrate maintenance problems, mechanical issues that may cause failure of CRAC units, and how such a failure can affect data center uptime, reliability and redundancy.

5. Basic understanding of the Liebert Site-Scan system, how it functions and monitors functions

6.	Understanding of the Ventilation system lay-out and functions	Demonstrate in-depth knowledge of hot-row/cold-row architecture, flow-paths of heat-waste to CRAC units, and strategies used to increase efficiency. Describe special requirements of EMC DMX cabinets, and steps to accommodate those requirements.
7.	Demonstrate a basic knowledge of the applicable sections of the OSHA standards and NFPA Codes	Describe EPA Section 608, discuss Clean Air Act, Montreal Protocol, Ozone depletion and causes. Describe risks of exposure to refrigerant, PPE. Discuss the "3 R's" of EPA Section 608, and their meaning as it pertains to the XXXXX Data Center.
8.	Demonstrate an understanding of tools and their uses	
9.	Demonstrate in-depth knowledge of the HVAC cooling strategies, and how they pertain to Data Center equipment and installations	Demonstrate in-depth knowledge of hot-row/cold-row architecture, power density limitations of XXXXX Data Center raised floor, and what can be done when those limitations are reached. Demonstrate operational awareness of future trends in IT equipment, and the ramifications for such trends, on data center uptime, reliability and redundancy.
10.	Shift heat loads from one CRAC to another	Perform manually, and using Site-scan.
11.	Assist and perform Preventative Maintenance on CRAC units	

SCDP CERTIFICATION FOR DUTY ENGINEER

SCDP Certification for Duty Engineer Signature Card Standard

D1. Abnormal/Casualty Operations Knowledge/Practical Factors:

1. Loss of utility electric power to the site
    Demonstrate in-depth understanding of different ways that utility power can be lost, and expected results.
    Heavy emphasis will be place on demonstration to "think on your feet," by using the one-line diagram and also by memory, without the assistance of the one-line diagram.

2. Loss of CRAC unit
    Demonstrate corrective actions needed on loss of CRAC unit.

3. Loss of UPS
    Demonstrate in-depth understanding of different ways that UPS power can be lost, and expected results.
    Heavy emphasis will be place on demonstration to "think on your feet," by using the one-line diagram and also by memory, without the assistance of the one-line diagram.

4. Fire in the Data Center
    Discuss immediate actions in the event of a fire. Demonstrate knowledge of EPO locations, how EPO system works, and ramifications of EPO operation.

5. Extreme inclement weather
    Describe impacts of extreme weather, how it affects the Pittsburgh Data Center

6. Loss of PDU
    Demonstrate in-depth understanding of different ways that PDU power can be lost, and expected results. Demonstrate immediate actions to take, if PDU is lost.
    Heavy emphasis will be place on demonstration to "think on your feet," by using the one-line diagram and also by memory, without the assistance of the one-line diagram.

7. Loss of Rack Power Strip
    Demonstrate corrective actions needed on loss of Power Strip.

8. Personnel Electrocution
    Demonstrate in-depth knowledge of electrical safety, and immediate actions to take if person

		shocked/electrocuted.
9.	Diesel Fuel Spill	Demonstrate in-depth knowledge of Hazmat requirements and immediate actions to take if a diesel fuel spill occurs.
10.	Battery Acid Spill	Demonstrate in-depth knowledge of Hazmat requirements and immediate actions to take if a battery acid spill occurs.
11.	Battery Thermal Runaway	Demonstrate knowledge of what a Thermal Runaway is, and the immediate actions to take, if it should happen.

SCDP CERTIFICATION FOR DUTY ENGINEER

SCDP Certification for Duty Engineer Signature Card Standard

D1. Abnormal/Casualty Operations Knowledge/Practical Factors:

12.	Failure of Utility Main Breakers	Demonstrate in-depth understanding of different ways that Utility Breakers may fail, and expected results. Demonstrate immediate actions to take, if a utility main breaker is lost. Heavy emphasis will be place on demonstration to "think on your feet," by using the one-line diagram and also by memory, without the assistance of the one-line diagram.
13.	Failure of 480 VAC Breakers	Demonstrate in-depth understanding of different ways that Load Center Breakers or critical bus breakers may fail, and expected results. Demonstrate immediate actions to take, if one of these breakers fails. Heavy emphasis will be place on demonstration to "think on your feet," by using the one-line diagram and also by memory, without the assistance of the one-line diagram.
14.	Loss of Switchgear Control Power	Demonstrate in-depth understanding of expected results of Switchgear Control Power failure. Demonstrate immediate actions to take, if this abnormal event should occur. Heavy emphasis will be place on demonstration to "think on your feet," by using the one-line diagram and also by memory, without the assistance of the one-line diagram.
15.	Failure of Switchgear Automatic Functions	Demonstrate in-depth understanding of expected results of Switchgear Automatic control failure (PLC). Demonstrate immediate actions to take, if this abnormal event should occur. Heavy emphasis will be place on demonstration to "think on your feet," by using the one-line diagram and also by memory, without the assistance of the one-line diagram.
16.	Failure of	Demonstrate in-depth understanding of

	4160/480 VAC Transformer	expected results of 4160/480 VAC Transformer failure. Demonstrate immediate actions to take, if this abnormal event should occur. Heavy emphasis will be place on demonstration to "think on your feet," by using the one-line diagram and also by memory, without the assistance of the one-line diagram.
17.	Failure of 480/120 VAC Transformer	Demonstrate in-depth understanding of expected results of 480/120 VAC Transformer failure. Demonstrate immediate actions to take, if this abnormal event should occur. Heavy emphasis will be place on demonstration to "think on your feet," by using the one-line diagram and also by memory, without the assistance of the one-line diagram.
18.	Failure of Emergency Generator During Emergency Generator Start/loss of power	Demonstrate in-depth understanding of different ways that Emergency Generator may fail, and expected results. Demonstrate immediate actions to take, if an Emergency Generator is lost. Heavy emphasis will be place on demonstration to "think on your feet," by using the one-line diagram and also by memory, without the assistance of the one-line diagram.
19.	Failure of Automatic Transfer Switch During loss of power	Demonstrate in-depth understanding of ramifications of a loss of ATS switches, and immediate actions to take. Heavy emphasis will be place on demonstration to "think on your feet," by using the one-line diagram and also by memory, without the assistance of the one-line diagram.
20.	Loss of one side of the electrical power distribution system	Demonstrate in-depth understanding of different ways that one side of the electric plant may be lost, and expected results. Demonstrate immediate actions to take, if an Emergency Generator is lost. Heavy emphasis will be place on demonstration to "think on your feet," by using the one-line diagram and also by memory, without the assistance of the one-line diagram.

SCDP CERTIFICATION RESPONSIBILITIES

APPENDIX D

SCDP CERTIFICATION RESPONSIBILITIES

Data Center Critical Operations Manager:

The Data Center Operations Manager is responsible for the following:

> 1. Implementation of the Skills Competencies Development Program (SCPD).
> 2. Periodic review of the program and administration of updates and modifications to the program.
> 3. Administration of the SCDP for Facilities Supervisor involvement and certification.
> 4. Review of the Duty Engineer candidate's progress and any remedial actions deemed necessary.
> 5. Recognition of the Final Certification of a candidate.

Facilities Supervisor:

The Facilities Supervisor is responsible for the following:
> 1. Initial evaluation of candidate to determine the candidate's skill, knowledge and ability levels.
> 2. Review and approval of candidate's Certification Programs.
> 3. Notification of candidates to establish an understanding of the Certification Program and to make candidates aware of their goals, what is expected of them, and to perform periodic evaluations of a candidate's progress.
> 4. Determination of a candidate's suitability prior to the end of the twenty-four week period given for completion of certification.
> 5. Administration of the SCDP for the Senior Maintenance Engineer's involvement and certification.
> 6. Establishment of a candidate's remedial actions if deemed necessary after an evaluation is done.
> 7. Ensuring that all authorized personnel authorized to sign a Certification Signature Card is qualified to do so.
> 8. Final Certification of a candidate.
> 9. Reporting of each candidate's progress in the Monthly

Operating Report, and direct reports to the Critical Operations Manager of any problems with a candidate's performance or progress.

Senior Maintenance Engineer (BTS):

The senior maintenance engineer is responsible for the following:

1. Aid in the evaluation of a candidate to determine the candidate's skill, knowledge and ability levels.

2. Perform periodic evaluations of a candidate's progress and keep them informed as to their identified strengths and weaknesses.

3. Recommendation of a candidate for Final Certification to support the Duty Engineer rotation.

Certified Duty Engineers:

Certified duty engineers are responsible for the following:

1.   Aid in evaluation of candidate to determine the candidate's skill, knowledge and ability levels.

2.   Forward recommendations to the Facilities Supervisor for a candidate's Certification and for establishing the manner of testing and evaluation to be done for the candidate.

3.   Perform periodic evaluations of a candidate's progress and keep them informed as to their identified strengths and weaknesses.

4.   Recommend candidates remedial actions and directions to accomplish them.

5.   Active involvement with training a candidate in the area that they are assigned cognizance for.

6.   Ensuring that all candidates are knowledgeable and capable before signing any Certification Signature.

SCDP CERTIFICATION FOR DUTY ENGINEER

APPENDIX E
SCDP CERTIFICATION FOR DUTY ENGINEER

SCDP FORMS

PURPOSE OF THIS APPENDIX

This appendix describes the forms that are used for documentation of SCDP progress for a candidate for Duty Engineer Certification. For each form there is an explanation of its purpose, followed by instructions for filling out each form. Included in the instructions are partially filled in forms, as examples, on how to apply the instructions for practical use.

SCDP FORMS

1.   PROGRESS GOALS/STATUS SUMMARY SHEET FOR DUTY ENGINEER: To establish the certification goals of a candidate for Duty Engineer. The Facilities Supervisor will outline the expected goals and flow-path for the candidate and the candidate will acknowledge these goals, on this form. This form will be retained for the duration of the candidate's certification training period by the Facilities Manager.

1.1.   This line reflects who the form is from. Typically, this will be the facilities supervisor to a new candidate, or to a Certified Duty Engineer that is required to do a bi-annual recertification.

1.2.   This line specifically names the individual that the form is targeted for. This will be the candidate for certification, or a certified Duty Engineer undergoing re-certification.

1.3.   The goals summary show the recommendations of the Senior Maintenance Engineer of the group prior to forwarding to the Facilities Supervisor.

1.4.   Here the Certification goals are listed out with attached start dates, expected completion dates, and any remarks deemed necessary by the facilities Supervisor, the Senior Maintenance Engineer, or any input from the candidate for which this is targeted.

1.5. The remarks section is provided for any information deemed pertinent, and/or useful to the staff and candidate that will benefit the candidate's performance.

1.6. Here the Senior Maintenance Engineer signs acknowledging the contents of the Goals/Summary Status Sheet.

1.7. Here the Facilities Supervisor signs; The Data Center Manager's signature here explicitly establishes this form as the primary goal for the candidate, or re-certifying Engineer, and delineates the specific course of action and performance required for the duration of the certification process.

1.8. The candidate signs the form acknowledging the candidate's personal goals and performance requirements for the duration of their certification period.

SCDP FORMS

PROGRESS GOALS/STATUS SUMMARY SHEET FOR DUTY ENGINEER CERTIFICATON

From (1.1) Facilities Supervisor

To: (1.2) John Newhire
Cc: (1.3) Senior Maintenance Engineer (BTS)

Subject: Expected Progress Sequence

    1.   You are expected to pursue certification by the Facilities Supervisor to support the Duty Engineer Rotation as outlined.

    2.   Your progress will be monitored periodically by the Facilities Supervisor and the Senior Maintenance Engineer. If you are unable to meet any of the assigned goal dates, you are to contact the Facilities Supervisor for guidance.

(1.4) CERTIFICATION GOALS	START DATE	EXPECTED COMPLETION DATE	REMARKS
Section 1	1/3/09	1/17/09	Reading Assignments
Sections 2A, 2B, 2C	1/3/09	2/14/09	General Policies, Basic Electrical and Mechanical
Section 1	1/3/09	3/27/09	Electrical Knowledge and Practical Factors
Section 1	1/3/09	3/27/09	Mechanical Knowledge and Practical Factors
Section 1	1/3/09	5/1/09	Abnormal and Casualty Operations
Section 1	1/3/09	5/15/09	Recommended Courses & Schools
Section 1	1/3/09	5/15/09	Certification Card Complete

Remarks (1.5)

    -   Courses & Schools to attend: BAS, CPR/AED, CFC,

ERIC WOODELL, D.Sc.

UPS, Switchgear, HVAC
Submitted:_____(1.6)_Senior Maintenance Engineer_____
                    Senior Maintenance Engineer   date
Approveed:_____(1.7)_Fac. Supv._____
                    Facilities Supervisor   date
Acknowledged:____(1.8)_John_____Newhire

                    Candidate   date

SCDP FORMS

2. RECORD OF REVIEW- DUTY ENGINEER CERTIFICATION

Purpose: To provide a documented record of review of a candidate's progress for Duty Engineer certification. This form will be filled out each time the candidate's progress, and certification card, is reviewed. This can be done at any time and must be filled out by the senior person performing the review and held in the candidate's training file.

> 2.1 The candidate's name is entered on this line to designate it as pertaining to that candidate for any review of any part of the candidates SCDP progress and records.
>
> 2.2 The reviewer will fill in his name, title, the date of review, and any remarks (concerns) that may arise.

SCDP FORMS

RECORD OF REVIEW – DUTY ENGINEER CERTIFICATION
CANDIDATE: (2.1) John Newhire

NAME OF REVIEWER	TITLE OF REVIEWER	DATE:	REMARKS
(2.2) Senior Maintenance Engineer	Sr. Maintenance Engineer	1/17/09	Sections 1 is Complete
Facilities Supervisor	Facilities Supervisor	1/17/09	Good progress in electrical and mechanical basics.
Senior Maintenance Engineer	Sr. Maintenance Engineer	1/24/09	Progress slowing down.
Senior Maintenance Engineer	Sr. Maintenance Engineer	1/31/09	Progress back up; progress is good…
Facilities Supervisor	Facilities Supervisor	1/31/09	On schedule.
Senior Maintenance Engineer	Sr. Maintenance Engineer		Section 1 is complete.

SCDP FORMS

3. <u>DUTY ENGINEER CERTIFICATION- PROGRESS RECORD</u>
Purpose: To document the progress of a candidate pursuing Duty Engineer certification. This form will be filled out at least weekly by the Facilities Supervisor or the senior maintenance engineer. This form should be used in conjunction with the Record of Review- Duty Engineer form to assess a candidate's progress towards Duty Engineer certification.

3.1   The candidate's name goes on this line.

3.2   The date the candidate started the certification process is entered here, below the candidate's name.

3.3   Each progress goal is listed and a numerical value is entered for each week, up to the date where certification is expected to be complete. This should be filled in by the Senior Maintenance Engineer, or the Facilities Supervisor. Notice that for the sake of example, there is a period where the imaginary candidate is performing poorly and an upgrading has been assigned to him.

3.4   The total number of signatures that the candidate has acquired to date is filled in on this line, above the total number of signatures that encompass the entire SCDP Certification Signature Card.

3.5   The percentage of completion for certification, for the week, is recorded here.

3.6   The required percentage of completion for that week is listed here.

3.7   The total percentage required for certification is recorded here.

3.8   The required percentage of completion to date is listed here.

3.9   Whether or not the candidate's progress is satisfactory or not is annotated here (i.e., SAT or UNSAT).

3.10   The Senior Maintenance Engineer reviews the candidate's progress and initials here for this review.

3.11   The Facilities Supervisor reviews the candidate's progress and initials here for this review.

3.12   The Facilities Supervisor initial here to acknowledge the progress of the candidate was entered into the MOR, at the appropriate time of the MOR submittal.

SCDP FORMS

DUTY ENGINEER CERTIFICATION- PROGRESS REVIEW

Candidate:_____ (3.1)John Newhire_____

Date Started:_ (3.2) 1/3/2009_____

Progress Goal	Percentage Complete											
	Wk 1	Wk 2	Wk 3	Wk 4	Wk 5	Wk 6	Wk 7	Wk 8	Wk 9	Wk 10	Wk 11	Wk 12
(3.3) Section 1	9/8	14/14										
Sections 2A,2B1,2	8/8	16/16	21/24	32/32	35/40	38/44	47/47					
Section 2B				1	2/4	3/9	13/14	21/19	—	—	—	—
Section 2C		3/1	5/3	6/5	7/7	8/9	11/11	13/13	—			
Section 2D									—	—	—	
Section 2E						2/1	3/2	3/3	—	—	—	—
Section 2F								1/1	—	—	—	—
Upgrade Program					50 %	100 %						
Total Signatures	17/13	33/13	40/13	53/13	58/13	75/130	88/13	99/13				
Signatures Earned	17	16	7	13	5	7	23	11				
Signatures Required	16	15	10	10	14	12	13	9				
% Earned this week	13	12	6	5	4	5	18	8				

% Required	12	12	8	6	11	9	8	7				
Total % as of this	13	25	31	41	45	58	68	76				
Required Total %	12	24	32	39	50	59	68	76				
Certificati on Status	SA T	SA T	SA T	SA T	U N-	U N-	SA T	SA T				
Sr. Maintena	S M	S M	S M	S M	S M	SM E	S M	S M				
Facilities Superviso	FS	FS	FS	FS	FS	FS	FS	FS				
Facilities Superviso				FS				FS				

SCDP FORMS

DUTY ENGINEER CERTIFICATION- PROGRESS REVIEW

Candidate: John Newhire

Date Started: 1/3/2009

Progress Goal	Percentage Complete											
	Wk 13	Wk 14	Wk 15	Wk 16	Wk 17	Wk 18	Wk 19	Wk 20	Wk 21	Wk 22	Wk 23	Wk 24
Total Signatures												
Signatures Earned this												
Signatures Required												
% Earned this week												
% Required this Week												

Total % as of this Week											
Required Total % as											
Certification Status											
Sr. Maintenance											
Facilities Supervisor											
Facilities Supervisor											

SCDP FORMS

4.   DUTY ENGINEER – PERSONNEL UPGRADE PROGRAM

Purpose: To develop a plan of action for a candidate that is falling behind the expected progress for Duty Engineer certification, to meet these goals and to aid a candidate in improving identified weak areas in knowledge or skill. This form is to be utilized as an aid in assisting a candidate to do this and will be filled out by the Facilities Supervisor, with the Senior Maintenance Engineer's input and concurrence.

> 5.1   Who this is coming from goes on this line. Usually this will be the facilities supervisor.

> 5.2   The candidate's name is filled in on this line.

> 5.3   If the upgrade was developed by someone other than the Facilities Supervisor, the preparer will fill in their title here. The upgrade program will be presented to the Facilities Supervisor by the preparer, prior to giving it to the candidate.

> 5.4   The exact weakness(es) are identified here. This is designed to allow a candidate the opportunity to become aware of their weakness(es) and correct them. Therefore it is of the utmost importance that the identified weakness(es) are accurately described within this form.

> 5.5   The means to allow the candidate to correct their weakness(es) need to be accurately detailed here. The object is to clearly outline a corrective avenue for the candidate, allowing them the quickest and most thorough upgrading possible, and permitting them to remain current with certification goals that are still being pursued.

> 5.6   The date that the upgrade must be completed by is recorded here.

> 5.7   The Senior Maintenance Engineer will review the upgrading plan before the Facilities Supervisor does, to check it for completeness and to ensure that it is adequate to fulfill the intended upgrade goals.

5.8   The Facilities Supervisor will review the upgrading plan and sign here to approve it for the candidate to follow.

5.9   The candidate will review the upgrading plan and sign here, acknowledging that there is an upgrading plan and that it will be followed.

SCDP FORMS

DUTY ENGINEER CERTIFICATION – PERSONNEL UPGRADE PRGRAM

From: (4.1) Fac. Sup.

To: (4.2) John Newhire

Cc: (4.3) Senior Maintenance Engineer

Subject: Assignment in Personnel Upgrading Program

1.   The level of training for your personnel certification has been determined to require upgrading due to (identify specific deficiencies):
___(4.4)_____

___Behind in Basic Electrical knowledge and electrical systems._____

_____

—

_____

—

_____

—

2. The specific requirements for your program are as follows: (4.7)
___A)_____Extra_____time_____during_____the_____day_____for
study._____
___B)____Time____with____Electrical____Engineer____for____tutoring____on____basic
concepts._____
___C) Study and develop an understanding of the UPS Modules with
the___   _____electrical
engineer._____

_____

—

3.   Upgrading   must   be   completed   by:
_(4.6)_____

4.   Upgrading program submitted by:(4.7) Sr. Maintenance Engineer
Date:2/3/09

Sr. Maintenance Engineer

5.   Upgrading   program   approved   by: __(4.8)   Fac.Sup.
_____Date:2/4/09
Facilities Supervisor

6.   Upgrading   program   Acknowledged   by: __(4.9)   John   Newhire_
Date:2/5/09
Candidate

SCDP FORMS

5. DUTY ENGINEER – RECORD OF EVOLUTIONS

Purpose: To document any evolutions or special procedures that a candidate may be involved in. This is to be utilized as a training aid to track what "hands on" experience a candidate actually receives while pursuing certification. This form may be filled out by the assigned Duty Engineer, the Senior Maintenance Engineer, or the Facilities Supervisor.

5.1   The candidate's name is filled in here.

5.2   The start date for the candidate to commence the certification process is listed here.

5.3   Any evolution that the candidate has a chance to actually perform, or is involved in, will be recorded here.

SCDP FORMS

DUTY ENGINEER CERTIFICATION – RECORD OF EVOLUTIONS

Candidate:___(5.1) John Newhire___   From:___(5.2) 1/3/09___
to_____

Evolution (5.3)	First	Second	Third	Fourth
PM UPS Module B-3	2/11/09			
PM UPS Standby				
PM on Emerg. Diesel	1/21/09	1/22/09	2/7/09	
Operational Test of EDG	1/23/09	1/23/09	2/13/09	2/13/09
PM CRAC unit	2/5/09	2/5/09	2/5/09	
Shift CRAC units via BAS	2/7/09	2/11/09		
Electric Plant Operational	2/14/09			

SCDP FORMS

6. <u>DUTY ENGINEER CERTIFICATION-ABNORMAL/CASUALTY OPERATIONS</u>

Purpose: To document a candidate's training on abnormal or casualty situations. This may include; actual response to an emergency situation, walkthrough training on different scenarios and equipment response, and discussions on how to operate the facilities equipment during emergencies or abnormal conditions.

6.1   The candidate's name is filled in here.

6.2   The start date for the candidate to commence the certification process is listed here.

6.3   Any abnormal condition or casualty that the candidate has a chance to acutely perform, or is involved in, will be recorded here.

SCDP FORMS

DUTY ENGINEER CERTIFICATION – ABNORMAL/CASUALTY OPERATIONS

Candidate: (6.1) John Newhire   From: (6.2) 1/3/09 TO: